# FIREPOWER 2030

# FIREPOWER 2030

## Major General P K CHAKRAVORTY (retd), vsm

**KW**
KNOWLEDGE WORLD

KW Publishers Pvt Ltd
New Delhi

*in association with*

Centre for Land Warfare Studies
New Delhi

The Centre for Land Warfare Studies (CLAWS), New Delhi, is an autonomous think tank dealing with contemporary issues of national security and conceptual aspects of land warfare, including conventional and sub-conventional conflicts and terrorism. CLAWS conducts research that is futuristic in outlook and policy-oriented in approach.

Centre for Land Warfare Studies
RPSO Complex, Parade Road, Delhi Cantt, New Delhi 110010
**Tel:** +91.11.25691308 **Fax:** +91.11.25692347
**Email:** landwarfare@gmail.com **Website:** www.claws.in

Disclaimer: The contents of this book are based on the analysis of materials accessed from open sources and are the personal views of the author. The contents, therefore, may not be quoted or cited as representing the views or policy of the Government of India, or Integrated Headquarters of MoD (Army), or the Centre for Land Warfare Studies.

ISBN  978-93-81904-80-0

# CONTENTS

# FOREWORD

The role of modern armed forces is to prevent conflict through deterrence and if it does break out, to fight and win – preferably on the adversary's territory. Future wars on the Indian subcontinent are likely to be limited wars. These are likely to spin out of ongoing conflicts like the over six-decade-old military standoff along the Line of Control (LoC) in Jammu and Kashmir (J&K), the Actual Ground Position Line (AGPL) on Saltoro Ridge, west of the Siachen Glacier conflict zone, and the proxy war being waged by Pakistan. Despite the ongoing rapprochement between India and China, a limited border conflict cannot be ruled out due to the unresolved territorial and boundary dispute and the yet-to-be demarcated Line of Actual Control (LAC). As the ongoing conflicts are mainly along land borders in the mountains, there is a very high probability that the next conventional conflict will again break out in the mountains.

The next conflict in the Indian context will be fought under a nuclear overhang. Serious attempts will be made to ensure that the conflict remains confined to the mountains, as a spillover to the plains may escalate out of control to nuclear exchanges. In any future war that the armed forces are called upon to fight in the mountains, gaining, occupying and holding territory and evicting the enemy from Indian territory occupied by him will continue to remain important military aims. Only massive asymmetries of firepower through a joint AirLand campaign can possibly achieve the desired military objectives. If proof was needed of this self-evident fact, it was amply provided during the 50-day Kargil conflict in 1999 where asymmetric artillery firepower had paved the way for victory.

From its original status as a 'supporting' arm, whose firepower 'neutralised' large areas of ground with dispersion that was inherent in its functioning, the artillery has now graduated to a full-fledged combat arm that dominates the battlefield with its accurate firepower that causes extensive destruction. In the post-Pokhran 1998 and post-Kargil 1999 scenarios on the Indian subcontinent, artillery is clearly seen to be a decisive arm, indeed, even a battle-winning one.

Firepower and manoeuvre are two sides of the same coin and both complement each other. However, in tactical situations in which either one lags behind due to the fog of war, the other must rise to the occasion and compensate if a favourable outcome is to be achieved. It is well known that future conventional wars on the Indian subcontinent will be fought under the nuclear shadow. Hence, it will be extremely risky to plan a battle that involves deep manoeuvre, particularly in the plains. In such a situation, favourable outcomes will be possible only through the massive application of artillery and aerially-delivered firepower. This major restriction on the manoeuvre component of military operations on land will lead to much greater emphasis having to be placed on firepower to achieve military aims and objectives.

Once a threat has been discerned, modern artillery firing 155 mm precision strike ammunition can be employed across the full frontage and depth of the battlefield to cause extensive damage and destruction to the enemy's forces. Today, laser-guided artillery shells can destroy bunkers, bridges and small buildings with a Single-Shot Kill Probability (SSKP) as high as 80 to 90 per cent. Targets that can be seen by the troops in contact with the enemy can be 'illuminated' by a laser beam by a ground-based artillery observer (or spotter) carrying a laser target designator. Those targets that are behind crest lines and on reverse slopes can be designated by an airborne artillery observer in an Army Aviation helicopter or even by an Unmanned Aerial Vehicle (UAV). Improved Conventional Munitions (ICM) shells carrying anti-personnel grenades and lethal 'air-burst' ammunition can be 'dispensed' over soft targets such as administrative bases, rations and fuel storage dumps, headquarters and rest areas. These have to be accurately directed using commando artillery observers or TV camera equipped UAVs to achieve the desired effect. Other force multipliers include gun locating radars for effective real-time counter-bombardment, UAVs equipped with TV cameras and suitable for high altitude operations for target acquisition, accurate target engagement and damage assessment, and powerful Integrated Observation Equipment (IOE), fitted with night vision devices for long-range target engagement by day and night.

In offensive operations on the future battlefield, the artillery will launch fire assaults or "attack by firepower" in conjunction with other combat echelons to shape the battlefield and, ultimately, create suitable conditions for the decisive defeat of the enemy. In fact, with the long reach of its missiles, rockets and medium guns, artillery firepower will systematically degrade the enemy's preparations for the attack from the concentration area onwards. The concentrated application of massed artillery firepower will disrupt the enemy's combat cohesion throughout the defensive battle. The Indian artillery will play an increasingly important role in the successful execution of integrated land-air operations on the modern battlefield by Suppression of the Enemy's Air Defence (SEAD) assets to enable own attack helicopters to operate freely and to also enable ground attack aircraft of the Indian Air Force (IAF) to launch strikes successfully.

With its ever-increasing range and lethality, the artillery is now capable of simultaneously fighting the contact, intermediate and deep battles. Its nuclear-tipped ballistic missiles such as the Agni series will guarantee India's nuclear deterrence. Its conventionally armed ballistic missiles such as the Prithvi and Prahaar and long-range rockets like the Smerch and Pinaka will influence the final outcome of a battle by striking deep. Unmanned Combat Air

Vehicles (UCAVs) are likely to join the arsenal soon. The utility of UCAVs has been amply proved during the recent conflicts in Afghanistan and Iraq and their entry will add a new dimension to the firepower punch of the artillery. In short, the integrated and synergetic application of artillery firepower at the point of decision will gradually but surely pave the way for victory and also help to reduce the Army's casualties. The artillery will be a co-equal partner with the manoeuvre arms in the successful execution of firepower and manoeuvre, provided it is equipped with modern guns and rocket launchers without any further delay.

Efforts are also underway to add ballistic as well as cruise missiles to the artillery arsenal. The BrahMos supersonic cruise missile (Mach 2.8 to 3.0), with a precision strike capability, very high kill energy and range of 290 km, is being inducted into the Army. A ceremonial induction function of the Block-I version was held in July 2007. Since then, the Block-II version has successfully completed trials. It is a versatile missile that can be launched from TATRA mobile launchers and silos on land, aircraft and ships and, perhaps, in the future, also from submarines. Fifty BrahMos missiles are expected to be produced every year. These terrain hugging missiles are virtually immune to counter-measures due to their high speed and very low radar cross-section and are far superior to subsonic cruise missiles like Pakistan's Babur cruise missile. The BrahMos will be a true force multiplier.

In the field of Command, Control, Communications, Computers, Intelligence, Information, Surveillance, and Reconnaissance (C4I2SR) too the artillery has made some progress. Good command and control networks combined with state-of-the-art Reconnaissance, Surveillance and Target Acquisition (RSTA) systems will enable the optimum utilisation of all available resources during times when hundreds of calls for artillery fire saturate the networks and resources are at a premium. The introduction into service of the Artillery Command and Control System (ACCCS) is only the beginning of capacity building for effects-based operations that will form the backbone of network-centric operations in the future. However, achieving such capabilities requires huge capital investments and the funds necessary need to be planned for as part of the modernisation process.

Hence, it is imperative that artillery modernisation is undertaken with alacrity so as to generate both qualitative and quantitative firepower asymmetries to achieve unassailable dominance on the future battlefield. Unfortunately, the modernisation plans of the artillery have stagnated for one reason or another for almost two decades. The civilian political leadership, the military hierarchy and the bureaucracy need to enhance their efforts to break the logjam and perhaps even take calculated risks with procurement procedures, if necessary, in view of the operational urgency.

It was with the aim of analysing the role and functions of the artillery and the tools necessary to achieve firepower supremacy on future battlefields that the Centre for Land Warfare Studies (CLAWS), New Delhi, invited Maj Gen P K Chakravorty, VSM (Retd), a well-known defence analyst, to focus on the key issues discussed above. As anticipated, he has done all that he had been asked to do and much more. This short book, entitled *Firepower 2030*, written by Gen Chakravorty, begins by analysing the constituents of firepower and the evolution of firepower technology; it then focusses on the technologies that are likely to emerge in the field of firepower by 2030, presents an Indian perspective and goes on to war-game the employment of firepower in possible future scenarios on the Indian subcontinent. Finally, it focusses on the present and future efforts towards modernisation. The author's erudite findings and deductions will help policy-makers to fine-tune their calculations.

The author deserves the highest compliments for his excellent effort. *Firepower 2030* is a welcome addition to the scarce literature available on the subject. It should be compulsory reading for younger officers who need to be schooled in the finer nuances of the application of asymmetric firepower and their commanders who need to better understand the battle winning potential of artillery firepower. And, it will be educative reading for those involved in the complexities of military modernisation.


Brig **Gurmeet Kanwal** (Retd)
former Director, CLAWS
May 10, 2013

# ABBREVIATIONS

ABM        Anti-Ballistic Missile
AEW&C   Airborne Early Warning and Control
ALH        Advanced Light Helicopter
AWACS   Airborne Warning and Control System
AMSL     Above Mean Sea Level
Cal          Calibre
$C^4I^2SR$   Command, Control, Communications, Computers, Information, Intelligence, Surveillance and Reconnaissance
DPP        Defence Procurement Procedure
DRDO      Defence Research and Development Organisation
DPSU      Defence Public Service Unit
FGFA      Fifth Generation Fighter Aircraft
ICBM      Intercontinental Ballistic Missile
LOR        Letter of Request
LOA        Letter of Acceptance
MARV     Manoeuvrable Reentry Vehicle
MGS        Mounted Gun System
MIRV      Multiple Independently Targetable Reentry Vehicle
NCW        Network-Centric Warfare
PSDA      Post Strike Damage Assessment
RFP         Request For Proposal
SP           Self-Propelled
TAR         Tibet Autonomous Region
TEC         Technical Evaluation Committee
TEL         Transport Erector Launcher
Tr            Tracked
UAS         Unmanned Aerial System
UAV         Unmanned Aerial Vehicle
UCAV      Unmanned Combat Aerial Vehicle

# 1

# CONSTITUENTS OF FIREPOWER

*Victorious warriors win first and then go to war, while defeated warriors go to war and then seek to win.*

— Sun Tzu

Firepower, as per the *Oxford Dictionary*, is the capacity to destroy, measured by the number and size of guns available. Warfare quintessentially comprises two ingredients: firepower and manoeuvre. Both complement each other and synergise to defeat the enemy. Manoeuvre entails moving to positions of advantage with dexterity to outwit the enemy. World War II and the Arab-Israeli Wars of 1967 and 1973 enabled skilful use of manoeuvre to defeat the enemy. However, future conventional conflicts would be influenced by the nuclear backdrop, bringing about constraints in time and space. Further, a large number of conflicts is visualised in mountainous regions which lack space for manoeuvre. Accordingly, firepower will be predominant in all conflicts of the current century.

The Revolution in Military Affairs (RMA) has resulted in modernisation of global weaponry and provision of a galaxy of surveillance and communication devices. This has compelled all countries to restructure their armed forces and transform their strategic thinking. Nations today are moving towards network-centric warfare. This essentially links the elements comprising the sensor, command elements and shooter to engage targets instantaneously. Therefore, the current focus is on precision standoff strikes in real-time. Firepower presently is undertaken from land, sea, air and submerged surfaces of the sea. Outer space is currently being used for surveillance. It is likely to become an area for deployment of weapons in the near future. Technologically, it would be practical to consider deployment of Anti-Satellite (ASAT) and Direct Energy Weapons (DEWs) in this region, with the developments taking place in this field. Firepower should be capable of breaking the enemy's will to fight. This would entail causing physical as well as psychological damage to attenuate his mental capabilities and set in a fear psychosis. The process of employing firepower entails the need for surveillance, which would lead us to reconnaissance of selected areas, thereby leading to acquisition of targets. These targets would, based on their importance, be degraded or destroyed. This would be ascertained by undertaking Post Strike Damage Assessment (PSDA) and based on the assessment, reengaged till the result is accomplished. In the present net-centric environment, this is possible through the application of Command, Control, Communications, Computers, Information, Intelligence, Surveillance and Reconnaissance ($C^4I^2SR$). The importance of firepower in the 21$^{st}$ century is supreme. Victory in any future conflict, in the current century, will be generated through asymmetries of firepower.

**Constituents**

The constituents of firepower are platforms with ammunition, which can deliver from five dimensions: land, air, surface of the sea, below the surface of the sea (hereafter referred to as sub-surface) and, possibly, in the times to come, outer space. These are small arms, guns, mortars, rockets of artillery, tanks, aircraft, missiles, Unmanned Combat Aerial Vehicles (UCAVs), armed helicopters, submarines and, in the future, possibly, stations in outer space. The ammunition, which is the payload, is the most important element of firepower. The various types of ammunition would constitute the conventional series, which would comprise high explosive, smoke, illuminating, armour piercing, high explosive squash head, fuel air explosive, cluster, precision, sensor fused, incendiary and propaganda. Apart from these, there is a strategic variety of ammunition which comprises Nuclear, Biological and Chemical (NBC). A combination of platforms with different types of ammunition results in devastating firepower which paves the way for victory.

The constituents enable us to undertake net-centric operations in a full spectrum conflict. The conflicts visualised could be land operations in a counter-insurgency situation against non-state actors, air-land operations in a counter-insurgency operation or a conventional conflict, air operations for counter-air, air defence, sea operations for sea control or denial, sea-land operations to undertake amphibious operations, air-sea operations between two opposing naval task forces and, possibly by 2030, operations involving outer space. In these conflicts, victory would be attained by ensuring asymmetries of firepower.

The systems used in the Army are distributed according to the role of the corresponding arm in battle. Small arms are with the infantry; tanks with the armoured corps, Armoured Personnel Carriers (APCs) with the mechanised infantry; guns, rockets, missiles, UCAVs, and surveillance equipment with the Regiment of Artillery and helicopters with Army Aviation. Engineers play a critical role in the mobility of firepower platforms and ammunition, The Corps of Signals provides a substantial proportion of $C^4I^2SR$ requirements. Systems in the Navy comprise frigates, destroyers, cruisers, corvettes, aircraft carriers, air defence ships, submarines, fighter aircrafts, maritime surveillance aircraft, helicopters, coastal guns and missiles. Air Force systems would include aircraft which are fighters and bombers as also armed helicopters, UCAVs, missiles for engaging a variety of targets and surveillance equipment like aerostats, Unmanned Aeriel Vehicles (UAVs), radars and $C^4I^2SR$

aircraft like the Airborne Warning and Control System (AWACS). These constituents would need to be modernised to undertake net-centric warfare in a full spectrum conflict. This would entail synergy of these constituents to ensure precise standoff strikes from long ranges with increased lethality. Further, Battlefield Transparency (BFT) is to be enhanced to ensure that everything tactically important is seen, and what is seen, can be hit in real-time.

The constituents of firepower would be used in operations which could be counter-insurgency, sub-conventional, limited, conventional, nuclear, chemical and biological. The operations would have to be effect-based, with possibly no collateral damage. Other aspects which characterise future operations are non-linearity, enhanced battle space and low threshold against the use of nuclear weapons. It is pertinent to note that more and more countries are developing nuclear weapons covertly to ensure their security. However, conventional conflicts can still take place between countries holding nuclear weapons as was observed during the clash on the Ussuri river between Russia and China, as also the Kargil conflict in 1999.

The constituents of firepower need to be optimally used to create asymmetries, thereby ensuring victory. This would be attained by seeking and identifying appropriate targets in the area of interest. Thereafter, the battlefield is to be shaped to one's own advantage by seamless engagements with the right combination of constituents. Further steps are to be taken to shield against enemy counter-measures. Finally, strikes would be needed to degrade, suppress and destroy, thereby breaking the enemy's will to fight. This would be possible with modernisation of our assets and correct application of resources.

# 2

# EVOLUTION OF TECHNOLOGY IN DEVELOPMENT OF FIREPOWER

*Power flows from the barrel of a gun*
— Mao Zedong

## Genesis

Technology has influenced firepower ever since the advent of war. Warfare initially was confined to stones, bows and arrows. The doctrines and method of fighting have been affected by technical advancement through the ages. It has often been debated that changes in doctrines have resulted in advancement of technology. This is at best partially true, as scientific developments cause evolution of technology which enables the "Revolution in Military Affairs" (RMA). As a matter of fact, both doctrines and technology keep complementing each other, resulting in mutual advancement. Historically the evolution of firepower can be traced to the invention of gunpowder by the Chinese in the 9th century. This technology was used by the Chinese military forces against the Mongols, when they invaded northern China. There is no record of when gunpowder was first used, therefore, it is difficult to correlate the events leading to its invention and passage of the knowledge to the Middle East onwards to Europe. Technologists, however, agree that it was invented by Chinese alchemists searching for an elixir of immortality.

Firepower made its entry into the battlefield through the use of cannons. The history of cannons spans several centuries. Xian Zhang, a Chinese poet composed a poem in 1341, "The Iron Cannon Affair". This poem describes the firepower of a cannonball fired from an eruptor, which could pierce the heart or belly of a man or horse, and cause numerous casualties.[1] The proto shells described in the battle of Huolongjing were probably the first to be used.[2] The Chinese also mounted over 3,000 bronze and iron cast cannons on the Great Wall of China, to defend themselves against their adversaries. The cannons were later used by both the Mongol conquerors and the Koreans. Chinese soldiers fighting under the Mongols appear to have used hand cannons in the battles of Manchuria in the year 1288. Evidence for this has been found at the archaeological sites where these operations were conducted.[3] Further, during the siege of Pyongyang in 1593, Chinese Ming troops used a variety of cannons to bombard an equally large Japanese Army. Despite both forces having similar numbers, the Japanese were defeated on a single day, due to the relative superiority of Chinese firepower.[4] Guns came into being when the Europeans adopted gunpowder. They bored an opening in a cylinder of metal and drilled another hole of extremely small diameter at the closed end. Thereafter they poured gunpowder and a tight fitting lead ball which was ignited by a flame torch and the resulting force projected the lead ball into the target area. Often, a slow burning rope was used to control the initiation. By the 15th century, cannons were made in a greater variety of lengths and diameters. The guiding principle was that the longer the barrel, the longer the range. By the end of the century, cannons were made more mobile. Wheeled gun carriages and trunnions became common and the invention of the limber facilitated the transportation of guns. It was during this period that the Mughal Emperor Babur fought the first battle of Panipat on April 21, 1526, against Sultan Ibrahim Lodi of Delhi. Lodi had a greater proportion of soldiers and elephants. However, he had no artillery. On the contrary, Babur had cannons which he skilfully used to cause consternation among the elephants, thereby breaking Ibrahim Lodi's will to fight, thus, paving the way to victory. This was skilful use of firepower in our country in the 16th century.

King Gustavus Adolphus of Sweden emphasised the use of light cannons and mobility in his armed forces. He stopped using the 12-pounder gun and used a lighter gun for the field artillery. He replaced the 12-pounder with the 4-pounder and the 9-pounder. These could be operated by three men, and pulled by 12 horses. Further, he introduced a special cartridge which had the powder and the shot which reduced the time for loading, thereby resulting in very high rates of fire. [5] He also pioneered the cargo shot, in which a canister was filled with musket balls which was very effective against troops in the open. He also organised his guns into batteries and used the artillery skilfully to decimate opposing troops. In 1631, at the Battle of Breitenfeld, Adolphus defeated Johann Tserclaes, Count of Tily, in Germany, by his dexterity in the use of firepower. The Swedes were outnumbered, but they battered the Germans by their very high rates of fire which inflicted severe casualties. The Germans were decimated and lost the battle. During this period, efforts were made to ensure that guns could be aimed to hit a target. Accordingly, the range was controlled by measuring the angle of elevation, using a gunner's quadrant. However, in the absence of sights, aiming lacked accuracy.[6]

The French Revolution and the subsequent Napoleonic wars revolutionised military strategy. Being a gunner, Napoleon used firepower skilfully in all his operations. Prior to his rise to power,

Inspector General Jean-Baptiste Varvette de Gribeauval made a few technological innovations. Artillery pieces were made with components, which enabled mass production. Gun carriages were built to a standard model; the mobility of the guns was improved by harnessing the horses in pairs, instead of moving in file, hard wood axles replaced heavy iron ones, and accuracy was improved by the introduction of tangent sights, which enabled a layer to sight the gun on the target.[7] French artillery, essentially comprised the Gribeauval guns which were 4, 6, 8, 12-pounders with 6 and 8-inch howitzers.[8] These guns were extremely light and to compare, the barrel of the British 12-pounder weighed 3,150 pounds and the remaining components 6,500 pounds. The Gribeauval 12-pounder barrels weighed 2,174 pounds and the remaining components 4,367 pounds. Since Napoleon insisted on speed in conducting his manoeuvres, these lighter cannons provided the flexibility he desired. In addition, his Army possessed vast quantities of mortars, furnace bombs, grape and canister shot which produced devastating effects at the target end. Napoleon used firepower effectively in all his battles. He concentrated his guns to blast a hole in the enemy's defences, to enable the infantry to break in, and thereafter the cavalry broke out. Firepower played an important role in the sea battles, with most ships containing anything from 50 to 100 guns. Napoleon's flagship *L'Orient*, with 120 guns, was the most heavily armed vessel in the world. Napoleon's final battle at Waterloo saw him use many more guns than the British or Prussians. As the battlefield was muddy, the guns would foul with the ground during recoil, resulting in slow rates of fire and the projectile shots were buried and did not ricochet, thereby causing less casualties. Due to these disadvantages, along with the Prussians playing a better role, the French lost this critical battle. Napoleon optimised the usage of firepower in all his battles, thereby breaking the enemy's will to fight.

The technological process of rifling, which entails casting spiral lines inside the barrel of a gun, was applied around 1855.

This provided spin to the projectile which improved accuracy. The Armstrong gun, invented by William George Armstrong, had rifling and was accurate at its maximum range. The projectile fired from the gun could reportedly pierce a ship's side and explode inside the vessel, causing total destruction of the target. The gun was procured by the British Army and the Duke of Cambridge was so impressed with the equipment, that he declared that it "could do everything but speak."[9] The superior guns with rifling enhanced firepower tremendously. During the Opium Wars against China in the 19th century, British battleships bombarded the coastal areas and fortifications from safe distances, away from the Chinese guns. It may be pertinent to note that the shortest war in recorded history, the Anglo-Zanzibar War of 1896 was brought to a swift conclusion by effective firepower from British battle ships.[10]

It would be of interest to trace the developments in another field of firepower: rockets and missiles. Their usage dates back to the 13th century. They were possibly used by the Chinese against the Mongols in1232. In the 18th century, Tipu Sultan developed rockets which he used during the Second Anglo-Mysore War in 1792, against the forces of the East India Company. Their effective usage resulted in 3,820 soldiers being taken prisoner. The rockets were deployed by special rocket brigades called *Kushoons*. These rockets were later reengineered by the British and used for a limited period.

## 19th Century and the World Wars

The elements that Napoleon and the Duke of Wellington manipulated at Waterloo were firepower and manoeuvre. In the 19th century, there were tremendous improvements in small arms. The rifle was provided with a magazine which raised its rate of fire, the propellant was smokeless, which provided concealment and the range was enhanced. The firepower of the infantry was stepped up by the Maxim type automatic machine-guns which were very effective against artillery guns placed in the firing line, compelling these artillery pieces to either dig down for protection, or adopt steel shields against bullets, or move back further to be out of range of the bullet. Wherever the ground permitted, guns were placed out of the range of the machine-gun. In such cases, artillery firings and counter-fire preceded small arms fire before an assault. However, if ridges intervened between the guns and targets, there was a need to find a method by which the gun could be fired at angles of elevation to cross crests. A breakthrough came with the French 75 mm gun in 1897. The principle adopted by the gun was the Quick Firing (QF) principle. The recoil system was upgraded to comprise a hydrostatic buffer and a recuperator. This enabled the barrel to return to its firing position after the round was fired. This was a technological milestone, as prior to this, guns had only buffer systems which absorbed the force of recoil aided by scotches, inclined planes and spades. The gun had a dial sight which enabled successive rounds to be fired at the required range and bearing. This enabled fire to be directed by an observation post officer using a telephone, while the guns remained hidden behind cover. These guns enabled firing in  direct and indirect mode as also they were used for predicted fire without ranging.[11]

The Japanese broke new ground in the use of their artillery. They handled their QF weapons skilfully in Manchuria against the Russians. On September 01, 1904, in the battle of Sha-ho, the Japanese deployed their guns on reverse slopes, which ensured they were not visible to the enemy or to their own infantry. In addition, they fired concentrations from reinforcing guns with which they communicated by telephones. Communication by telephones enabled control of artillery fire which provided the optimisation of indirect firepower, which added to its versatility.[12] The war in Manchuria led to numerous debates about the employment of artillery. While some felt that the Japanese method of indirect fire from concealed positions was the appropriate method, there were others who felt that it was better to use guns in line with the infantry, firing at targets by direct firing. The problem in keeping guns in concealed positions on reverse slopes was mainly that of reliable communications. At this juncture, the only means available was the telephone which was prone to disruption due to cables being broken by the shelling. There was definitely a need to improve communications and have observation post officers at the target end and command post staff at the gun end.

World War I commenced in 1914, with the Germans moving into France realising the full potential of indirect fire being synergised with the infantry and cavalry. However, the French and the British, using their machine guns and artillery, were able to stall their advance. Both sides dug down and the entire situation turned into a stalemate, known as trench warfare.

The tank was developed to break the stalemate caused by trench warfare. The British Army tested the first prototype on September 08, 1915. To maintain secrecy, they were initially referred to as water carriers and thereafter termed as land ships. The British took the lead in tank development and were closely followed by the French, who fielded their first tanks in 1917, but the Germans were slower in this field. The first use of tanks on the battlefield was that of 49 British Mark-I tanks at the Battle of Somme on September 15, 1916, with mixed results as many broke down but a third succeeded in breaking through. The numerical tally was that out of the 49 tanks, 32 participated in the attack and nine made it across to "no man's land" to the German lines. The tanks had been rushed into combat, but their usage gave essential feedback of the modifications to be made. The French first used their tanks on April 16, 1917, during the Nivelle offensive. However, their tanks were destroyed by effective use of long range artillery, by the Germans. The first really successful use of tanks came in the Battle of Cambrai in 1917. Col Fuller, of the British Army with his tanks made an excellent break-in, but the break-out was done by horse cavalry, which marginally exploited the situation. By 1918, the British had produced the Mark-V tank and the French had produced a light tank weighing 8 tons. Further, the guns fitted on the tank were shortened to ensure they were stable while negotiating obstacles. The German General Staff were keener on anti-tank weapons than tanks. The only German tank to be developed was the A7B, of which 15 were manufactured. The first tank versus tank battle took place at Villers-Bretonneux on April 24, 1918. It was an unexpected engagement between three German tanks and three British tanks. The British tanks proved to be superior, but, the British plan of using mass armour could not fructify as the blockade of Germany and entry of the US brought an end to the war.[13] The tank gun had a higher muzzle velocity and could be gainfully employed in engaging Armoured Fighting Vehicles (AFVs) and field fortifications; its firepower produced devastating effects on the battlefield. World War I also saw the introduction of tethered balloons and aeroplanes for gathering information and for directing own artillery fire. These fixed-wing aircraft did a commendable job in directing firepower.

There was no doubt regarding the versatility of the tank in future conflicts. The period between the two World Wars saw tremendous improvement in the tank by the British, French and Germans. It was during this period that the self-propelled gun, which was needed to move as a part of the mechanised column was developed and inducted. The period witnessed the development of aircraft and weapon systems. Further, the Germans were developing the V-1 buzz bomb and the V-2 stratospheric rocket. Between the two wars, aircraft technology improved, resulting in production of aircraft which were capable of providing firepower in air-to-air, air-to-land and air-to-sea engagements. Further, technological developments led to the production of the 3.7 inch howitzer which could be split into separate components and carried by mules. In addition, efforts were being made to tow guns with motorised vehicles. The 25-pounder was under development. The Germans, despite the World War I armistice requirements, developed tanks, aircraft, towed artillery guns and mortars. The Americans replaced the 75 mm gun of World War I with the 105 mm howitzer. Further, heavier models, including the 155 mm "Long Tom" gun and 8-inch howitzer were developed. They also focussed on communications. Radio sets were introduced to control fire. The communications set-up enabled a forward observer with an infantry company to maintain constant contact with his guns to his rear. Fire commands from observers were received by Fire Direction Centres (FDCs) located with gun batteries. The FDCs used map grids, firing tables and instruments to compute the aiming data necessary to hit an unseen target. An officer in each FDC had the communications necessary to bring all guns within range from a division or a corps to engage a target.[14] Further effort was made to improve the variety of ammunition by the European countries and the USA. Development of air power and its capability of influencing ground battles were enhanced with the invention of the jet engine by Sir Frank Whittle of the UK in 1930 and Hans von Ohain of Germany in 1936. The German jet was flown in August 1939 and the first British jet was flown in May 1941. These were to influence battles dramatically in the times ahead as the jets with their high speeds could provide firepower to the ground forces which would pulverise enemy field defences.

World War II witnessed massed usage of tanks, self-propelled artillery, rockets, guns and mortars. Gen Guderian, the Corps Commander who led the *blitzkrieg* stated that the most important aspect of a tank was its firepower; protection and mobility were characteristics which were essential but merited less importance.

It is pertinent to note that during the war, operations of the land forces were effectively combined offensive use of air power. Technologically, firepower in this war was also influenced by use of armed aircraft at Pearl Harbour, V-1 flying bombs, V- 2 rockets and nuclear weapons in Hiroshima and Nagasaki.

The attack on Pearl Harbour by the Japanese was a surprise military strike against the USA on the morning of December 07, 1941. The attack was carried out with the aim of preventing the US Pacific fleet from interfering with the planned Japanese operations into Malaya and the Dutch East Indies. On November 26, 1941, a Japanese task force of six aircraft carriers left northern Japan enroute to a position northwest of Hawaii. From this area, they launched 408 aircraft for the mission. Out of these, 360 aircraft were used for the two attack waves and 48 were used for defensive combat air patrol. The attack resulted in damage of all eight US Navy battleships, three cruisers, one mine-layer, three destroyers and an anti-aircraft ship. Apart from this, 188 aircraft were destroyed on the ground, 2,402 Americans were killed and 1,282 were wounded. The attack saw the formal entry of the USA in World War II and demonstrated the devastating effects of massed firepower.

The V-1 flying bomb was developed by the German Air Force during World War II. The weapon was launched at London one week after the successful landing of the Allies in Europe on June 13, 1944. The V-1 guidance system used a simple auto pilot to regulate altitude and air speed. There was a sophisticated interaction among the yaw, pitch and roll. Several bombs were provided with radio transmitters to check the general direction of flight and the target's grid coordinates by radio bearing. An odometer driven by a vane anemometer on the nose determined the arrival on the target area to undertake area bombing. The V-1s had a range of 96 km and could be air as well as ground launched. Some 9,521 V-1 bombs were used against London and 2,448 against Antwerp in Belgium. The attacks caused 22,892 casualties, most of whom were civilians. It is of interest to recall how these bombs were countered. The average speed of the V-1 was 560 km per hour and it flew at an altitude of 910 m (3,000 ft). The British used barrage balloons and fighter aircraft: 2,000 barrage balloons were deployed and 300 V-1s were destroyed by these balloons. The aircraft used were the Tempest, Mosquito, Spitfire XIV and Mustang.

The V-2 rocket was a ballistic missile that was used mainly for targeting London and Antwerp. This was the world's first long range ballistic missile and the first known man-made object to enter space. Over 3,000 V-2 rockets were launched, resulting in the death of about 7,250 military personnel and civilians. The rocket ranged about 350 km and reached an altitude of 100 to 110 km. It had four major technological advances: its powerful engine, aerodynamic shape, innovative guidance system and radio transmission system. The rocket engine was fuelled by an ethanol and water combination, with liquid oxygen serving as an oxidiser. The rocket was 14m long, 1.5 m in diameter, weighed 20 tons and had a warhead that weighed 907 kg. The rocket was guided by an Inertial Navigation System (INS) which was relatively inaccurate and, therefore, unsuitable against military targets. Most of the scientists involved in the project migrated to the USA or USSR after the war.

Firepower in its nuclear version was used during World War II, resulting in the surrender of Japan, accepted on August 14, 1945, and signed on September 02, 1945. The weapon was developed by the Manhattan Project under the direction of Maj Gen Leslie Groves, of the US Army Corps of Engineers. Two types of bombs were eventually manufactured at Los Alamos under American physicist Robert Oppenheimer. The Hiroshima bomb, known as Little Boy, was a fission weapon used with uranium-235, a rare isotope of uranium extracted in factories located at Oak Ridge, Tennessee. The other, at Nagasaki, known as Fat Man, was an implosion type of nuclear weapon using plutonium-239, a synthetic element created in nuclear reactors at Hanford, Washington. Hiroshima was bombed on August 06, 1945. The 393 Bomber Squadron provided the B-29 aircraft nicknamed Enola Gay. Enola Gay was accompanied by two other B-29s. They were launched from Northfield air base on Tinian in the West Pacific. The estimated flying time was six hours. A reconnaissance mission signalled to the bomber that the weather was perfect and Enola Gay arrived over the target in clear visibility. The bomb was released at 8.15 a.m. local time. It took 43 seconds to fall from the aircraft which was at a height of 31,060 ft. The pre-detonation height was about 1,900 ft above the city. Due to cross-winds, it missed the aiming point, the Aioi bridge by almost 240 m and detonated directly over a surgical clinic. The blast equivalent was about 13 kilotons of TNT. The radius of total destruction was about 1.6 km, with resulting fires across 11 sq km. Japanese officials determined that 69 percent of the buildings were destroyed, 6 to 7 percent were damaged, 70,000 to 80,000 people were killed immediately and another 70,000 injured. Over 90 percent of the doctors and 93 percent of the nurses were casualties.

The second bomb was dropped over Nagasasaki, a port city, at 11.01 a.m. on August 09, 1945. The blast equivalent was 21 tons of TNT. Casualties reported were extremely high. Firepower, produced by the nuclear weapons broke the will of the Japanese and compelled them to accept the terms of surrender on August 14, 1945.[15]

### Post-World War II

The Korean War, fought in the 1950s, saw the introduction of the helicopter, but it was only used for ferrying troops and was not weaponised. It was only during the Vietnam War, which saw US involvement commence in 1964, that an armed helicopter was introduced. The Huey AH-1B attack helicopter was used extensively to carry out search and destroy missions in Vietnam.

The US Army created its first full scale air mobile unit, the 11$^{th}$ Air Assault Division, in 1964. This was commanded by Brig Gen Harry W.O. Kinnard. It was a conventional light division with 434 helicopters. The Huey Cobra helicopters of the two light helicopter battalions provided the lift to carry infantry. Firepower was provided by three conventional artillery battalions that could be lifted by Chinook helicopters and an aerial artillery battalion consisting of rocket firing Hueys. Gen Kinnard built and employed his division in an unconventional manner. Having undergone its trials on July 01, 1965, the division was redesignated as the 1$^{st}$ Cavalry Division Airmobile. Two months after its activation, the division arrived in Vietnam. It was pitched by Gen Westmoreland against North Vietnamese Military Region IV in the central highlands. Gen Kinnard, by audacious use of firepower, was able to destroy a large portion of the three North Vietnamese regiments that attacked the base at Plei Me and used, for the first time, artillery deployed by helicopters on the ground and Huey helicopters firing rockets from the air. Apart from this, Forward Air Controllers (FACs) called in fighters that performed creditably by both day and night. The use of effective firepower resulted in tremendous success during the operations conducted from October 27, 1965 to November 15, 1965.[16] The Vietnam War also introduced the UAV which later got upgraded to UCAV and precision ammunition. As this was a war fought under asymmetric conditions, technological developments were focussed on being able to fly and fight in all weather conditions by day and night.

The UAV which was used in Vietnam for surveillance and directing artillery fire soon became a force multiplier in numerous operations. During the Lebanon crisis in 1982, the Israeli UAVs completely deceived the air defence system of the Syrians and were able to effectively direct Israeli firepower on the Syrian forces. The UAVs graduated to UCAVs and revolutionised firepower. Their potential as platforms for suppression of enemy air defence systems, early warning of ballistic missile attacks and even boost phase intercept of ballistic missile makes them flexible and effective, able to undertake engagements at low cost. They can achieve these results without risking the lives of trained pilots and also pave the way for victory by destruction without collateral damage. UCAVs have already demonstrated their ability in counter-insurgency operations. A number of Al Qaeda leaders have been killed in precision attacks by UCAVs. The US UCAVs have mounted Lockheed Martin's Hellfire missile, whereas Israeli UCAVs have the mounted Lahat, a semi-active laser guided missile with a range of 10 km.

Precision ammunition is guided to accurately hit a target with minimum collateral damage. The electro-optical bomb was used during the Vietnam War. The entire equipment comprised a television camera and flare sights by which the bomb would be steered until the flare superimposed the target. The camera bombs transmitted a bomb's eye view of the target to the controlling aircraft. Simultaneously, development took place in the field of laser guided weapons. All these weapons rely on the target being illuminated by a laser target designator on the ground or an aerial platform. The laser designator sends its beam in a coded series of pulses to ensure that the bomb is not confused by other signals. The best use of these bombs was by the US Air Force on April 27 and May 13, 1972 against the Thanh Hoa bridge in Northern Vietnam. This bridge had been the target of 800 US Air Force sorties using normal bombs, but the same was successfully destroyed by Phantom jets using Laser Guided Bombs (LGBs). The use of technology in the use of firepower can be effectively compared by comparing two important battles, Dien Bien Phu and Khe Sanh. At Dien Bien Phu, the Vietnamese defeated the French, and at Khe Sanh, troops of the 26th Marine Regiment held on due to the fusion of electronic and conventional firepower. A brief comparison of these two battles would clearly highlight this aspect.

| Factors | Dien Bien Phu | Khe Sanh |
|---|---|---|
| Air lines of communication | 272 km from Hanoi | Helicopters, located at 56 km; tactical aircraft at 144 km |
| Vietnamese communication from China | Long | Short |
| Defenders | 13,000 French | 6,000 Marines |
| Key outposts | Fell early | Held |
| Attackers | 50,000 Vietnamese | 20,000 Vietnamese |
| Defending artillery | 24 light guns | 18 light howitzers, 6 medium howitzers and 24 heavy guns |
| Attacking artillery | 200 light and medium guns | Total unknown 100, 122, 130, 152 |
| Factors | Dien Bien Phu | Khe Sanh |
| Attacking artillery | multiple rocket launchers | mm guns and (continued) howitzers,140 mm rockets |
| Average rounds (per day) | 2000+ | 150 |
| TAC air (per day) | 30 to 40 sorties with less than 200 French aircraft | 377 sorties 2,000 airplanes, 3,300 helicopters |
| Total ammunition | French less than 2,000 tons | 110,000 tons |
| Defenders' casualties | 2,700 killed, 4,400 wounded | 205 killed, 1,668 wounded |
| Vietnamese casualties | 7,900 killed, 15,000 wounded | 10,000-15,000 killed |
| Result | French defeated | Marines held on to the position |

The asymmetry of firepower means of a higher technology enabled the Marines to hold on to Khe Sanh. The battle for Khe Sanh lasted for 77 days, which was about three weeks more than Dien Bien Phu. Khe Sanh held due to two reasons; first the traditional tenacity of the Marines and, second, the availability of an effective firepower system that combined, for the first time in warfare, both the electronic and conventional ammunition in high magnitude.[17] High technology, as also the fusion of conventional firepower and electronic means was the difference at Khe Sanh where the Marines held on despite numerous attacks by the Vietnamese

Post Vietnam, development of precision weapons gained tremendous importance as collateral damage became unacceptable. In the field of artillery guns, the 155 mm Cannon Launched Guided Projectile (CLGP) was the first to be developed. This was referred to as the 'Copperhead', which was a 155 mm fin stabilised terminally guided laser projectile. The projectile would function once the target was illuminated with a laser. The projectile would detect the laser signal, activating the onboard guidance system and operate the steering vanes to manoeuvre the projectile onto the target. The range of the projectile was a minimum of 3 km and a maximum of 16 km. The Russians developed a similar system known as Krasnopol which was used for the 152 mm and 155 mm gun systems. These systems were effective against tanks and hard targets. However, they needed good visibility and a lot depended on the skill of the observer operating the laser designator. Accordingly, two types of precision ammunition were developed for the artillery: the precision guided kit and the extended range precision guided shells. There are two known precision guided kits in use: the first is the Alliant Tech system XM 1156; and the second, the Israeli Top Gun GPS/ INS 2 D course correction fuse. Both are screwed into the nose of the existing projectile like the existing fuse. The XM 1156 is guided by the Global Positioning System (GPS) and has an accuracy of 30 m, whereas the Top Gun is guided by INS with GPS in the loop and provides an accuracy of 20m. The M982 Excalibur is a 155 mm extended range artillery shell developed by Raytheon Missile System and BAE System. With a combination of the GPS satellite and inertial guidance, the round provides an accuracy of 4 to 6 m at a range of 37 km. The Block II of this ammunition carries either 65 Dual Purpose Improved Conventional Munitions (DPICMs) or two Sense and Destroy Armour Munitions (SADARMs). The Excalibur can be fired with the 155 mm gun and has been used operationally in Iraq and Afghanistan.

In the category of precision ammunition, we also have the sensor fused ammunition which can be fired from the 155 mm gun. This is a carrier shell containing two cylindrical sub-munitions. It has an electronic nose fuse on which the time is set before firing. The fuse is set to function at an altitude of over 1,000 m above the target area. At the appropriate time, a small expelling charge ejects two container cylinders, each containing one sub-munition, from the rear of the projectile body. Hereafter, each container cylinder is dispersed, while the rotation rate and velocity are reduced by spin brakes. After a fixed time, a secondary ejection operation separates each sub-munition from its container cylinder. As each sub-munition descends, a stabilising disc is released, after which two wings and the Electro-Optical Unit (EOU) fold out with a spin rate of 15 revolutions per second. In the case of the Bonus, ammunition wings were selected in lieu of a parachute, as they are less sensitive to strong winds and provide a smooth search pattern. The EOU is equipped with a multi-band passive Infrared (IR) detector or a mm wave radar which is switched on at the appropriate time, and homes on to a target. The sensor fused ammunition's sensitivity makes it a potent weapon against mechanised vehicles.

Another area of technological development comprises cluster munitions. A cluster munition is a form of air-dropped or ground-launched explosive weapon that releases or ejects smaller munition. This variety of ammunition has an enhanced lethal area that destroys personnel, vehicles, runways, electric power transmission lines and propaganda leaflets. Many of the unexploded bombs can kill or injure civilians after a conflict has ended and are difficult to locate and remove. Cluster bombs could be anti-personnel, anti-tank, anti-runway, mines, anti-electrical, leaf dispensing, incendiary and chemical weapons. Anti-personnel cluster bombs use explosive fragmentation to cause casualties to personnel and destroy soft targets. They were widely used during the Vietnam War, when vast numbers of cluster bombs were dropped over Vietnam, Laos and Cambodia. The anti-tank variety in most cases contains shaped charged warheads to pierce the armour of tanks and Armoured Personnel Carriers (APCs). Modern guided sub-munitions such as those found in the US CBU-97, can use either a shaped charge or an explosively formed penetrator. Unguided shaped charge cluster ammunition is designed to be effective against entrenchments with overhead cover. Anti-runway cluster bombs are designed to penetrate concrete before detonating, allowing them to shatter and crater runway surfaces. The British JP 233 uses a two-stage warhead that combines a shaped charge as a conventional bulk explosive charge. The shaped charge creates a small crater which is expanded by the detonation of a bulk explosive charge. Anti-runway cluster bombs are normally used along with anti-personnel cluster bombs equipped with delay or booby trap fuses that make repair work difficult. Cluster bombs are used to dispense mines and these do not detonate immediately but behave like conventional mines. Mines manufactured by the USA are designed for self-destruction after 4 to 48 hours, to keep the battlefield clear of explosives. The anti-electric cluster bomb, the CBU-94/B, was first used by the USA in the Kosovo War in 1999. The bomb consists of a tactical ammunition dispenser filled with 202 BLU-114/ bomblets. Each bomblet contains a small explosive charge that disperses 147 reels of fine conductive fibre of either carbon or aluminium coated glass. Their purpose is to disrupt and damage electric power transmissions by producing short circuits in high voltage power lines and electric sub-stations. On the first attack, these knocked out 70 percent of the electric power supply in Serbia. The same ammunition was used effectively against Iraq in Gulf War II. Reports indicate that it took 500 people 15 hours to get one transformer yard to get back on line after being hit with these weapons. The leaflet cluster bombs are used for propaganda purposes. In this connection, the LBU-30 bomb with leaflets has been tested by an F-16 aircraft flying at 20,000 ft in the year 2000. The LBU-30 consists of SUU-30 dispensers that have been adapted to leaflet dispersal. The dispensers are usually recycled units from old bombs. It is important to note that enclosing the leaflets within the bomblets ensures that leaflets will fall on the intended area without being dispersed excessively by the wind. Cluster weapons to be used for delivery of chemical payloads were developed by the United States and the Soviet Union during the period from 1950 to 1960. The Chemical Weapons Convention of 1993 banned their use. Thereafter, these cluster bombs were destroyed. The last in the cluster bomb series is the incendiary bomb. These bombs are intended to start fires and have sub-munitions of white phosphorus or napalm, and they often include anti-personnel and anti-tank submunitions to hamper fire-fighting efforts. The incendiary bomb technologically resulted in the development of the thermobaric bomb.

Technological developments in the evolution of firepower have witnessed the development of the powerful thermobaric bomb. This is an explosive that produces a blast wave of longer duration, thereby increasing the number of infrastructural and human casualties. These explosives rely on atmospheric oxygen, where most explosives have a fuel oxidiser premix (gunpowder contains 25 percent fuel and 75 percent oxidiser). They have significant advantages when deployed inside confined environments such as tunnels, caves and bunkers. Thermobaric weapons have been used by the Russians and the US forces. On March 03, 2002, the US used a single 2,000 lb bomb laser guided thermobaric bomb against cave complexes in which Al Qaeda and Taliban fighters had taken refuge in the Gardez region of Afghanistan. The effect was devastating and blew up the entire target into smithereens.

## Missiles

During World War II, the US and USSR both started research programmes on rockets and missiles based on the German V-2 design. A variety of missiles was manufactured, from the Short Range Ballistic Missile (SRBM) onwards to the Intermediate Range Ballistic Missile (IRBM) and finally the Intercontinental Ballistic Missile (ICBM). The USSR manufactured the first ICBM in 1957 and the USA in 1960. The flight path enabled them to launch many space systems. The flight phases comprise the boost phase, the mid-course phase and the reentry phase. The boost phase lasts for about three to five minutes. The missile altitude at the end of this phase is typically 150 to 400 km, depending on the trajectory chosen. The mid-course phase takes approximately 25 minutes. During this phase, it is in an elliptical flight path. The apogee half way through the elliptical path is at an altitude of approximately 1,200 km. During this phase, the missile continues on an unpowered ballistic trajectory like an artillery projectile. The reentry phase commences at an altitude of 100 km and lasts for two minutes with an impact speed of 4 km per second. ICBMs can be deployed from multiple platforms, which could be missile silos, submarines, trucks and mobile launchers on rails. Modern ICBMs carry Multiple Independently Targetable Reentry Vehicles (MIRVs) each of which carries a separate nuclear warhead allowing a single missile to hit multiple targets. The MIRVs are effective against an Anti-Ballistic System (ABM) and are currently held by the US, Russia, China, France, Britain and possibly Israel.

While the ballistic missiles in most cases carried a nuclear payload, cruise missiles which flew at lower altitudes and could be steered to the target with precision, have been developed and have been used extensively in the Gulf Wars, Afghanistan, Kosovo and Libya. The most extensively used cruise missile has been the US subsonic Tomahawk missile with ranges of about 1,700 km and guided by GPS, INS, TERCOM (Terrain Contour Matching) and DSMAC (Digital Scene Matching Area Correlation). The

Tomahawks were mainly launched from naval ships, but a few were launched from the F-117. The missile is accurate and its effects are devastating. The Indo-Russian joint venture cruise missile developed a decade ago is a fire and forget cruise missile which has a range of 290 km and is a precision weapon capable of undertaking surgical strikes. This is a unique system with platforms on land, sea, and is in the process of development for the air version.

Precision technology in the field of missiles has resulted in the development of loitering missiles in which the missile has the sensor and the shooter located in the missile. Accordingly, the missile can select a target and engage it with pinpoint accuracy. Further, its endurance enables the missile to loiter over the target area, thereby providing it the opportunity to select and engage a target. The endurance of a loitering missile varies from 30 minutes to 8 hours. The missile is in its final stages of development and would be inducted into service thereafter. Once inducted, the missile would be ideal against targets such as bridges, railway siding, logistic areas, critical points in a military station, headworks, missile sites, ammunition dumps and command and control centres. It may be pertinent to note that another precise missile is the short range fire and forget Hellfire missile. The weapon is an air-to-surface and surface-to-surface missile. It has a high explosive anti-tank metal augmented charge, a solid fuel rocket engine and an operational range of 500 m to 8 km. The guidance system is the millimetre wave radar seeker. It has been launched from rotary and fixed-wing platforms as also UCAVs, tripods, ships and ground vehicles. The missile killed Hamas leader Ahmed Yassin in 2004 and Anwar al Awlaki, Al Qaeda leader, in 2011.

Technology has also seen the development of rocket systems by Russia and the US. Though these systems are not very accurate, they produce shock action by delivering large quantities of ammunition simultaneously on the target. The GRAD BM-

21 and Smerch of Russia; the MLRS and HIMARS of the US; and the Pinaka weapon system produced by India are systems which cause shock by their awesome firepower. The Smerch has a maximum range of 90 km and fires a variety of ammunition which can pulverise a target, thereby breaking the adversery's will to fight.

## Anti-Ballistic Missile

An Anti-Ballistic Missile (ABM) counters ballistic missiles, thereby enhancing the firepower of own missile system. The term includes any anti-missile system designed to counter ballistic missiles. Many short range tactical ABM systems are currently operationalised. The best known systems are the US Army Patriot, US Navy Aegis and Israeli Arrow. The Patriot missile performed correctly during the Gulf War. Russia has the S-300, S-400 and S-500 ABMs. China has acquired the S-300 system and is developing other ABMs. India has also developed ABMs which have been successfully test-fired. Against ICBMs, there are only two systems — the Russian A-135 (the Gorgon and Gazelle) and the US ground-based mid-course defence. The interceptors are based in Alaska and provide protection against missiles launched from North Korea, China and Russia. ABMs are not successful against MIRVs and it is economically ineffective to develop weapon systems to counter these multiple warheads. With the withdrawal of the US from the ABM Treaty in 2002, the Russians are not complying with START II, thereby maintaining the MIRVs against which no ABM system is practicable.

Recently, the Israeli Defence Forces (IDF) have developed and inducted the IronDome mobile air defence system. This has been developed by the Rafael Advanced Defence System and is designed to intercept artillery rounds and rockets. The system has three main components, a detection-cum-tracking radar, a Battle Management Weapon Control Centre (BMC) and a missile firing unit that launches the Tamir interceptor missile equipped with electro-optic sensors as also several steering fins for high manoeuvrability. The radar detects a round or a rocket launch and tracks its trajectory. Thereafter, the BMC calculates the expected point of impact and fires the interceptor missile to destroy the rocket over a safe area. The launch platform has three launchers, each carrying 20 interceptors, The system has been deployed by Israel in early 2011 and has been effective in the Gaza Strip against rocket attacks.

## Directed Energy Weapon

This is a new class of weapon system, currently under development. This weapon emits energy in an aimed direction without the means of a projectile. It transfers energy to a target for desired effect. The energy can come in the form of electromagnetic radiation in lasers or masers, particle beam weapons, flame throwers and sound in sonic weapons. Laser weapons usually generate brief high energy pulses. Most existing weaponised lasers are gas dynamic lasers. They use a low powered oscillator to generate a coherent wave which can destroy a target. The pulsed energy projectile emits an infrared laser pulse which creates rapidly expanding plasma at the target. The resulting sound, shock and electromagnetic waves stun the target and cause temporary pain and paralysis. Further, there are particle beam weapons that can use charged or neutral particles to decimate targets. The last weapon in this category is a sonic weapon. These weapons utilise sound to injure, incapacitate or kill an opponent. Currently, they are in the development phase. These are in the form of sonic bullets, sonic grenades, sonic mines or sonic cannons. They make a focussed beam of sound and some make sound over an extensive area. Sound waves of high power emitted by these weapons can disrupt eardrums and cause severe pain or disorientation. On the contrary, less powerful sound waves can cause nausea and discomfort. These weapons would be optimised once the development phase is completed.

## Command, Control, Communications, Computers, Information, Intelligence, Surveillance and Reconnaissance ($C^4 I^2 SR$)

Evolution of technology has changed the dimensions of war from human-centric to platform-centric and currently the move is towards Network-Centric Warfare (NCW). This warfare integrates sensors, decision-makers and firepower. The entire process in the current battlefield milieu has integrated these elements, resulting in synergy during operations. This has been possible due to the development of multifarious sensors, communication links between sensors and command elements as also with weapons to engage selected targets and PSDA to know the efficacy of engagement.

Broadly, there is a wide array of surveillance equipment. Sensors are located on land, sea, under water, in the air and outer space. Ground-based surveillance equipment comprises the Thermal Imaging Integrated Observation Equipment (TIIOE), Long Range Observation and Reconnaissance Equipment (LORROS), short range battlefield surveillance radars, medium range battlefield surveillance radars, weapon locating radars and sound ranging system. Naval surveillance equipment comprises radars, seabed arrays and sonar. Surveillance from the air comprises aerostats, AWACs, tactical reconnaissance, UAVs and UCAVs. Space-based systems comprise geo-synchronous satellites, remote sensing satellites, low earth orbiting satellites and space stations. Information provided by these sensors is interpreted by the command and control elements leading to decisions regarding target engagements. A tactical $C^4 I^2 SR$ system would be primarily based on the Command Information Decision Support System (CIDSS) with affiliated systems comprising the Artillery Combat Command and Control System (ACCCS), Electronic Warfare System (EWS), Battlefield Surveillance System (BSS), Air Defence Control & Reporting System (AD C&RS), Operational Air Support (OAS), Air Space Control System (ASCS) and

Communications and Data Network System (CDNS). These systems enable the data to be processed, leading to selection and engagement of targets thereby resulting in effective firepower at the objectives to accomplish the operational aim.

Evolution of technology has enhanced our firepower potential by great magnitude. The present battlefield environment provides good surveillance, creditable reconnaissance leading to systematic target acquisition followed by engagements with an aim to degrade or destroy, thereby breaking the enemy's will to fight and paving the way to victory.

## Notes

1. John Norris, *Early Gunpowder Artillery: 1300-1600* (Marlborough: The Crowood Press, 2003), p.11.

2. Joseph Needham, *Science and Civilisation in China*, Vol 7, *The Gunpowder Epic* (Cambridge University Press, 1987), pp. 263-275.

3. Arnold Pacey, *Technology in World Civilisation: A Thousand-Year History* (MIT Press, 1990), p. 47.

4. Christon I Archer, *World History of Warfare* (University of Nebraska, 2002), p.211.

5. Albert Manucy, *Artillery Through the Ages* (Diane Publishing, 1994), pp. 7-8.

6. Tunis Edwin, *Weapons: A Pictorial History* (John Hopkins University Press, 1999), p. 9.

7. Richard Podruchny, *The Success of Napoleon* (Military History on line.com), p.1.

8. Owen Connelly, *Blundering to Glory: Napoleon's Military Campaigns*, Third Edition (Rowman & Littlefield Publishers, 1979), Introduction, p.xii.

9. Marshall J Bastable, *Arms and the State: Sir William Armstrong and the Remaking of British Naval Power* (Ashgate Publishing Limited, 2004), pp.72-73 and 94.

10. Mark C Young, *Guinness Book of World Records* (Bantam Books, 2002), p.112.

11. Shelford Bidwell and Domnick Graham, *Fire Power: British Army Weapons and Theories of War, 1904-1945* (Hampstead, UK: George Allen and Unwin Publishers Limited,), pp. 7-10.

12. Captain B.Vincent, "Artillery in the Manchurian Campaign," *Royal United*

*Services Institution Journal,* Vol 52, 1908.

13. John Glanfield, excerpts from *Devil's Chariots: The Birth and Secret Battles of the First Tanks* (Sutton Publisher, 2006).

14. Robert H Scales, Jr, *Firepower in Limited War* (Washington D.C.: National Defence University Press, April 1990), p.11.

15. President Truman's Papers, "Hiroshima, US Strategic Bombing Survey. The Effects of Atomic Bombing of Hiroshima and Nagasaki", pp. 11 to 16.

16. Scales, Jr, n.14, pp. 21-22 and 66-74.

17. Ibid.

# 3

# TECHNOLOGICAL DEVELOPMENTS BY 2030: IMPACT ON FIREPOWER

*The speed, accuracy and devastating power of American Artillery won confidence and admiration from the troops it supported and inspired fear and respect of the enemy.*

— Gen Dwight D Eisenhower

## Anticipated Developments

Prolific growth of technology would take place by 2030 and this would have a profound impact on firepower and its applications. The technologies of robotics, nano, directed energy, thermal imaging, modular charge system, millimetre wave, photonics, smart materials, cyber and space-based weapon systems are in their development phase and would lead to stealth, light weight, enhanced range and precision.

These technologies would make firepower all-weather, more pervasive and extremely devastating. Nano technology would make platforms smaller and lighter, enabling greater carriage and application, particularly in the mountains. The ranges of artillery gun systems would be further enhanced by the use of the Velocity Enhanced Long Range Projectile (VLAP) and

Vulcano ammunition. The VLAP ammunition is presently being developed in South Africa and currently has a range of 60 km with a 155 mm, 52 calibre gun. VLAP technology combines base drag reduction and rocket propulsion. The VLAP projectile shares an identical external interface with a 155 mm projectile. The Vulcano ammunition is primarily sub-calibre ammunition which would range to about 100 km. As far as the gun barrel length is concerned, we have reached a finite length of 52 calibres. Further increase in barrel length will affect stability and mobility. There is a case for the introduction of soft recoil like the one used in the 105 mm XM 204 which would reduce the length of recoil, with recoil being controlled by the forward movement of the barrel before firing, thereby making the weapon more stable and accurate. Modular and bi-modular charges have already been introduced; superior charge systems will be much more efficient, leading to less erosion of the barrel. Apart from this, there is a need to develop precision ammunition for surgical strikes. This is extremely important in engagement for counter-insurgency tasks where terminally-guided mortar bombs would be of tremendous use against militant camps. Robotics needs to be exploited to the hilt in the light of the success of the UCAV MQ-9 Reaper over the Afghanistan-Pakistan region. The UCAVs with their Hellfire missiles have been able to precisely knock out targets in most cases. There were some civilian casualties, to avoid which would need better recognition and identification of targets. Possibly, we would be soon having UCAVs with air-to-air missiles and direct energy weapons. The Pentagon plans to expand its global network of drones and special operations in a fundamental realignment meant to project US power even if it cuts back on conventional forces. The plan which Defence Secretary Leon Panetta unveiled in early 2012 calls for a 30 percent increase in the US fleet of UCAVs in the coming years. It also sees the deployment of more special operations teams at a growing number of small bases.[1]

Precision for any weapon system would be based on the Inertial Navigation System (INS) which would be corrected by a satellite navigation system, followed ultimately by a terminal guidance system which could either be a seeker or be designated to the target by laser. While the INS and the seeker or designator is controlled within the weapon system, navigation from the satellites is presently through the US Global Positioning System (GPS) or the Russian system, Glonass. While we have signed a precision code agreement with Russia, giving us an accuracy of two metres, there is a definite need to have our own navigation system. In the immediate future, we will have a GPS-based Geo-Augmented Navigation System (GAGAN). This is a planned implementation of a regional Satellite-Based Augmentation System (SBAS) by the Indian government. It is a system to improve the accuracy of a Global Navigation Satellite System (GNSS) receiver by providing reference signals. The project involves establishment of a full complement of SBAS consisting of 15 Indian reference stations (INRES), 3 Indian Navigation Land Uplink Stations (INLUS), 3 Indian Mission Control Centres (INMCC), 3 geo-stationary navigational payloads in C and L bands with all the associated software and communication links. The GPS is a satellite navigation system designed to provide instantaneous position, velocity and time information anywhere on the globe. The base line satellite constellation consists of 24 satellites positioned in six earth centred orbital planes. The orbital period of a GPS satellite is one half of a sidereal day or 11 hours 58 minutes. The orbits are nearly circular and equally spaced about the Equator at a 60 degree separation, with an inclination of 55 degrees relative to the Equator. The orbital radius is approximately 26,000 km. With the base line satellite constellation, users with a clear view of the sky have a minimum of four satellites in view. The current GPS constellation cannot support requirements for all phases of flight, and integrity is not guaranteed. All satellites are not monitored at all times; the time-to-time alarm is from minutes to hours and there is no indication of quality of service. In the GAGAN system, the GNSS data is received and processed at widely dispersed INRES which are strategically located to provide coverage over the required service area. Data is forwarded to the INMCC, which processes the data from multiple INRES to determine the differential corrections and residual errors for each monitored satellite and for each predetermined Ionospheric Grid Point (IGP).

Information from the INMCC is sent to the INLUS and uplinked along with the GEO navigation message to the GAGAN satellite. The GAGAN satellite downlinks this data to the users via two L band ranging signal frequencies, L1 and L5. This improves the accuracy of our GPS-based system. While GAGAN would be an intermediate step for better coordinates from the US GPS, there is a need for our own navigational system, as in the US, Russia, European Union and China. At the current pace, GAGAN would stabilise in another six years and thereafter, the focus must shift to creation of an Indian navigation system. For missiles to be effective in the conventional mode, the Circular Error of Probability (CEP) must be less so as to ensure that the target is effectively destroyed. To reduce the CEP, it would be essential to obtain the precision code for the GPS or Glonass. While Russia has expressed its willingness to offer the precision code, its effectiveness has to be tested. In the case of the US military, accuracies would be possible only if we possibly sign agreements related to communications which would not be in our strategic interest. Either way, we have to have our independent navigation system which should be able to provide us real-time data to guide our weapon systems. The India Space Research Organisation (ISRO), along with the Ministry of Defence, has to work out the details of the system and make it operational within a decade. This is imperative as China has its own navigation system.

Stealth would play an extremely important part in all weapon systems to ensure that they reach points from which they can effectively release their weapons. Stealth is obtained by altering the radar shape, using radar absorbing paints and, lastly, smart material which provides incorrect reflexes to the radar. Stealth would be used for all types of platforms and missiles. These would enable freedom of movement which would cause force multiplication of firepower at the target end. Thermal imaging in all our observation equipment would be enhanced with observation equipment from the air being able to reach realistic ranges of 80 km. By 2030, our air assets would include the Fifth Generation Fighter Aircraft (FGFA) as also Longbow Apache helicopters with fire-and-forget anti-tank guided missiles which could be either the Nag or an advanced form of the Hellfire missile. The Defence Research and Development Organisation (DRDO) and the Original Equipment Manufacturers (OEMs) of our UAVs are developing weaponised payloads for converting existing and future UAVs into UCAVs.

By 2030, India with assistance from Russia would have inducted the FGFA. The signed contract between India's Hindustan Aeronautics Limited (HAL) and Russia's United Aircraft Corporation (UAC) stipulates that 214 fighters will be built for India and 250 for Russia. Although there is no reliable information about the specifications as yet, it is known from sources that the FGFA will be stealthy and would be outfitted with the next generation of air-to-air, air-to-surface, air-to-ship missiles and incorporate an Active Electronically Scanned Array (AESA) radar which can track 32 targets and engage eight of them in near simultaneity. The armament comprises two 30 mm internal cannons and 16 hard points (eight internal and eight external) for mounting Beyond Visual Range (BVR) surface-to-air or surface-to-land missiles. The aircraft is expected to have a speed of Mach 2 and is comparable to the US F-22 Raptor as also the F-35 Lightning II and the Chinese Chengdu J-20.

Simple flight demonstrations of the aircraft have been carried out by Russia in 2012 and development will be completed possibly by the end of 2014. It would possibly commence induction in the Indian Air Force (IAF) around 2017, and we would have reasonable number of aircraft by 2030.[2]

By 2030, India would have stabilised its ICBM system with missiles ranging to possibly 10,000 km. It is also expected that the MIRVs would have been developed and inducted in our Strategic Forces. Further the K-15 Submarine Launched Ballistic Missile (SLBM) would have been optimised for our submarines with possibly ranges greater than the present range of about 700 km. Prahaar, a solid fuelled surface-to-surface guided short range tactical ballistic missile with omni-directional warheads and a range of 150 km, was test-fired successfully on July 21, 2011. The missile has a disadvantage in engaging targets with conventional warheads as it adopts a ballistic path which would lead the adversary to believe that the weapon in flight could be a nuclear-tipped missile, thereby inviting a possible asymmetric nuclear response. Therefore, the engagement of targets by conventional warheads in our environment would generally be confined to cruise missiles. In this field, Nirbhay, a subsonic cruise missile with a range of 1,000 km, would have been developed and inducted. In about two years, the cruise missile, BrahMos, would be inducted for air-to-land and subsurface-to-sea or land versions. Further, the BrahMos, with hyper-sonic speed, would possibly have been developed and inducted into the armed forces in about seven years. There is a need for a lighter cruise missile which could be carried by a UCAV and be launched from the air on ground and sea targets. Apart from this, the anti-tank guided missile, the Nag, would have met the user requirements and been inducted into the armed forces. Further, the Beyond Visual Range Air-to-Air Missile (BVRAAM) Astra which is an 80-km class, active radar-guided missile meant for beyond visual range air-to-air combat, is currently under development, and is likely to be inducted on completion of user trials in another five years. The ABM project of DRDO unveiled in 2006 has two missiles, the Advanced Air Defence (AAD) and the Prithvi Air Defence (PAD) missiles. The AAD is an endo-atmospheric interceptor of new design which can intercept targets to a height of 30 km. The PAD is a modified Prithvi missile which is an exo-atmospheric interceptor up to an altitude of 80 km. In the field of rockets, there would be an increase in the range of the Smerch and Pinaka weapon systems. The Smerch, with a new rocket, would range up to 120 km, and the Pinaka would develop a rocket that would attain maximum range of 60 km. A lighter Smerch vehicle based on the Kamaz vehicle, possibly carrying eight tubes, has been developed and would be available for use in mountainous terrain. This would enhance capabilities to bring greater volume of firepower in the mountains.

Contrary to the Outer Space Convention, gradual weaponisation of space is taking place. In January 2001, a US Congress Space Commission headed by Mr Donald Rumsfeld, who later became the Secretary of Defence, recommended that the US government should vigorously pursue the capabilities called for in the National Space Policy to ensure that the country will have the option to deploy weapons in space to deter adversaries and defend US interests.[3] Developments are complete for Anti-Satellite (ASAT) weapons to be launched from the ground, as also preparations are on for launching them from space, when required. The US visualises space dominance by a combination of ASAT weapons as also a layered missile defence system. The missile defence system would provide the capability to engage ballistic missiles in all phases of flight soon after they are launched, at the vertex height, and as they descend in the terminal phase. This is being done through space-based kinetic energy interceptors and advanced target tracking satellites. Experiments on the interceptor are expected to commence in 2012. Development is secretly being undertaken of Low Earth Orbit (LEO), with its first near field infrared experiment satellite designed to gather information on ballistic missiles during the first few minutes of their flight. This could be imaginatively used as the interceptor. Further research on the space-based laser has been conducted for destroying missiles in the initial phase after being launched. There are other space programmes which can be turned into weapons. The US Air Force has a research programme of an Experimental Satellite (XSS) that seeks to use satellites to conduct proximity operations. These would be used for destruction of other satellites and could also be used to inspect and service friendly satellites. The US Air Force is also considering acquiring weapons for global force projection from outer space such as the aero vehicle and hyper velocity rod bundles. These weapons could enable any target to be engaged in 90 minutes on the earth's surface, thereby providing tremendous flexibility to the user. Currently, all these are in the development phase, and a gradual weaponisation of space is taking place. This weaponisation would enhance the firepower capabilities of a nation exponentially.[4]

Direct energy weapons which are currently under development would be inducted in small numbers by 2030. $C^4I^2SR$ would be optimised and full spectrum operations would be undertaken in a network-centric environment. This would enhance the effectiveness of firepower.

## Impact of Technological Developments on Firepower

A full spectrum conflict by 2030 will have the total impact of technological developments which would enable engagements in the network-centric environment to be undertaken with speed and precision. The aim would be to attack by firepower in all stages of the battle so as to achieve favourable conditions for the decisive defeat of the enemy. Firepower would be effectively used against non-state actors in a sub-conventional conflict and against adversaries in a conventional war. During a sub-conventional conflict, firepower will have to be moderated based on the proportion of force needed to be applied to avoid collateral damage. Precision would be needed to ensure that targets are correctly addressed and thereafter destroyed. Actionable intelligence is extremely important in discerning a target and thereafter engaging it in real-time to achieve decisive results. The platforms used for such engagements would vary from aircraft, attack helicopters, cruise missiles, ballistic missiles, short range terminally guided missiles, guns with terminally guided ammunition, UCAVs, rocket launchers and small arms with holographic sights. Firepower on such occasions would have to be delivered with a high degree of accuracy by day and night. Missions will be undertaken by teams that should be able to exploit the technical capabilities of the weapons and ammunition. Special Forces, equipped with blue tooth communications, light weight laser designators, third generation image intensifying night vision equipment, along with light weight thermal image integrated operating equipment with the capability to direct fire from aircraft, UCAVs and artillery guns, would be able to handle targets effectively in such conflicts. Technology combined with grit and determination, accompanied by good actionable intelligence would enable precise firepower to decimate conventional targets.

Firepower will continue to play a predominant role in conventional conflicts against a nuclear backdrop in 2030. Territorial disputes, particularly in Asia, could result in inter-state disputes, leading to war. In such scenarios, the role of all fire assets would be applied judiciously at critical points to physically and psychologically degrade the cohesion of the enemy, with the ultimate aim of breaking his will to fight. As space for manoeuvre will be restricted, firepower will be applied at the strategic, operational and tactical planes. At the strategic plane, the aim would be to target the enemy's war-waging potential. Engagement of the enemy, while undertaking speedy mobilisation will prevent him from building favourable force ratios. This would be undertaken by applying firepower on economic centres, infrastructure, military and industrial bases as also strategic reserves. At the operational plane, firepower would impose unfavourable force correlation by decimating the centre of gravity of the enemy's forces. The enemy's freedom of action in the battle zone would be curtailed by degrading his command and control network. In addition, asymmetry in firepower would enable the attacker to degrade enemy artillery and other fire assets. At the tactical levels, firepower would be primarily focussed on the destruction of enemy defences and troops in open and field fortifications. Thereafter, application would be on firepower delivery means, reserves and counter-attack forces. Optimisation of firepower would be achieved by judicious application of air, attack helicopters, guns, mortars, rockets and electronic warfare assets against the enemy in contact or where contact is imminent. Overall, firepower in 2030 would pulverise the enemy on the objective, thereby facilitating ease of capture. As far as India is concerned, we will not be able to produce all types of state-of-the-art weapons system by 2030 and, therefore, will have to rely on import of a few weapon systems.

## Notes

1. Adam Entous, Julian E Barnes and Shobhan Gorman, "America's Military Plans Realignment," *The Wall Street Journal*, January 27-29, 2012.
2. Rajat Pandit, "India-Russia to Ink New Military Pact," Times News Network, Times of India.indiatimes.com, October10, 2009.
3. "Report to the Commission to Assess US National Security Space Management and Organisation," Washington DC, January 11, 2011.
4. US Air Force, "Counterspace Operations,"Air Force Doctrine, pp.2-2.1.

# 4

# INDIAN PERSPECTIVE

*Artillery conquers and infantry occupies.*
— Maj Gen J F C Fuller

## Introduction

India has to contend with both state and non-state actors. It has hostile borders with its western neighbour, Pakistan, and northern neighbour, China. It has to contend with sub-conventional conflicts in Jammu and Kashmir (J&K) and the northeastern states. Further, the Naxalite problem has been stated by Prime Minister Manmohan Singh to be the single biggest internal security challenge that India has ever faced. The movement has affected life in over 200 districts in the states of West Bengal, Bihar, Andhra Pradesh, Jharkhand, Odisha, Chhattisgarh, Madhya Pradesh and Maharashtra.[1] Accordingly, non-state actors are involved in insurgent activities and hostile borders exist with China and Pakistan. India has numerous island territories, oil assets in the South China Sea and there is a threat to commercial ships from pirates operating in the Indian Ocean. To effectively deal with these problems, there would be a need for effective surveillance, target acquisition and, if the need arises, to engage with a view to degrade or destroy these targets. Surveillance, target acquisition and engagement form the three important constituents of firepower and there is a need to examine their application in the complex Indian environment.

## Sub-conventional Conflicts

Sub-conventional conflicts with non-state actors are being undertaken in J&K, the northeastern states and with the Naxalites in the states of West Bengal, Jharkhand, Bihar, Odisha, Madhya Pradesh, Chhattisgarh and Andhra Pradesh. Sub-conventional threats today are more relevant to national security than ever before. The nation faces multifaceted challenges from these militants who are today rightly termed as the biggest threat to our country and, if not dealt with effectively, could turn out to be an extremely difficult problem to resolve. While there is a need for political measures to improve the general situation, there is a need to undertake surveillance, carry out tactical assessment launch operations and use firepower to decimate the militants. In J&K, all the militant groups are actively supported by Pakistan. The main terrorist groups currently operating are the Lashkar-e-Tayyeba (LeT), Jaish-e-Mohammed (JeM) and the Harkat-ul-Mujahideen (HuM). They are driven by a radicalist Islamic agenda rather than any negotiable grievances. Of these, the LeT is the most formidable and is capable of sustained operations even out of South Asia.[2] There are about 42 training camps in Pakistan Occupied Kashmir (POK) where the militants are trained and are thereafter launched for undertaking terrorist operations in the state. These training camps which are also launch pads, have to be kept under surveillance by satellites that must give inputs of dispositions as also any shift in locations of these camps. If the need arises, we should be able to bring down effective firepower by the use of cruise missiles, fighter aircraft in the ground attack role, attack helicopters, UCAVs and medium range artillery to destroy these camps. The Heron UAVs could undertake PSDA and correct as also adjust fire wherever applicable. Use of precision ammunition would minimise collateral damage, thereby, possibly avoiding a large scale conflagration. These militants normally are inducted prior to the melting of the snow and our surveillance devices like the LORROS, THOE, N Cross, other night vision devices and the UAVs would be able to detect them and use suitable ambushes to surprise them and bring down accurate small arms fire to effectively terminate their mission. Troops deployed on the Line of Control (LoC) would undertake this task with dexterity and imagination. The residual militants may manage to get into the villages and other local habitats. In such a situation, there is the need for accurate information obtained through human intelligence, regarding their location. Thereafter cordon and search operations may be undertaken by the security forces, in which small arms are used. There is a scope of using mortars for such operations provided they can fire precision ammunition as also if confirmed information about the militants is available, so that innocent civilians are not killed. This should be done with great care and possibly when the militants are away from inhabited areas, in their jungle hideouts. By 2030, precision ammunition would have been developed and this would enable precise engagement, particularly in militant hideouts under thick foliage, away from villages and built up areas. Further, use of propaganda leaflets may be needed in remote areas. The railway line between Jammu-Udhampur-Katra-Banihal-Quadigunj-Srinagar-Baramulla would have been completed and would need surveillance as also protection by firepower, combined with mobile patrols. The situation would have progressively improved but would still need military handling due to its proximity to Pakistan and China.

Numerous groups are involved in the insurgency in northeast India. The region comprises seven states (also known as the seven sisters): Assam, Meghalaya, Tripura, Arunachal Pradesh, Mizoram, Manipur and Nagaland. Regional tensions have eased of late, with concerted efforts by the government to raise the living standards of the people in the region. However, insurgency to a lesser degree continues in Assam, Manipur, Nagaland and Tripura. The insurgents in these states have been trained either in Bangladesh or Myanmar, and are less aggressive in comparison to those of Jammu and Kashmir. Most of the actions need small arms; however, mortars with precision ammunition could play an important role in the future on selected occasions. Surveillance with UAVs would be practical using high resolution Synthetic Aperture Radars (SARs). These would enable observation under thick cover. It is to India's advantage that Bhutan, Bangladesh and Myanmar are now not permitting these militant groups to train and operate in their countries. Overall, the situation would possibly stabilise by 2030, subject to development being undertaken at a steady pace in all these states.

The Naxalite issue has so far been dealt with by the police authorities, with training support from the Army. Gradually, the police forces are coming to terms with the problem and are able to improve their surveillance and overall handling of the problem. UAVs with SARs would be able to provide essential inputs which could be acted upon by the security forces. Small arms along with cautious use of mortars, would be ideal for these tasks. A resolution of the Naxal problem would need socio-economic improvement and this would take a long time, viewing its slow pace.

**Full Spectrum Conflict**
The recently released Chinese military budget, in March 2012, entails a sum of US $ 106 billion. This is practically three times the Indian defence budget and focusses on a stronger People's Liberation Army (PLA), with emphasis on jointness, information warfare, and development of firepower assets, naval craft and stealth fighters. China's seventh White Paper, released in March 2011, states a three-fold challenge to China: Taiwan, East Turkestan and Tibet. Despite Chinese protests, the US continues to supply arms and spares to the Taiwanese forces, thereby ruling

out brute force as a viable alternative. With regard to the Uighur movement, in East Turkestan, China has been ruthless in quelling the protests. In connection with Tibet, China has left no stone unturned to impose the Han culture on the autonomous region. Further, it has openly staked a claim to Arunachal Pradesh and even objected to the visit of India's Defence Minister Shri A K Antony to the state in February 2012. There are numerous incursions by the PLA all along the Line of Actual Control (LAC) which are progressively increasing, despite deliberations. Further, China has objected to Vietnam giving two oil blocks to the Oil and Natural Gas Commission (ONGC) Videsh in the South China Sea. The only thing booming between the two countries is commercial trade which currently is in the region of $ 74 billion. The last war that China fought was in 1979, which was to teach a lesson to Vietnam. Maj Gen Shelford Bidwell, a British military analyst, has stated that the Chinese have conceptualised a new form of war known as the "teach a lesson" model. He is of the opinion that the Sino-Indian War of 1962 was the first campaign in this category. The war was intended to teach India a lesson for supporting the Dalai Lama. The dispute along the borders was a cause which was cunningly used to launch a surprise attack, thereby effecting a defeat. This was followed by a voluntary withdrawal that demonstrated the inability of India to respond militarily to the crisis. The Chinese applied the same model to Vietnam but they did not succeed in the operations and had to realise the hard way, the need for modernising their forces and moving from people's war to war under high-tech conditions.

The doctrine enunciated by the Chinese remains germane as it limits a conflict and prevents its escalation. Accordingly, it enables a limited conflict to be undertaken against a nuclear backdrop. The historical fact is that it was effective in the Soviet era and prevented any escalation of the Sino-Vietnam War to a Sino-Soviet War. The teach a lesson doctrine could be easily used by China vis-a-vis India, where a sudden conventional limited war is waged against a nuclear backdrop. It is pertinent to note that China has also warned India not to interfere in the disputed waters of the South China Sea. China has numerous issues of disagreement with India. Therefore, in an authoritarian set-up, to fight a limited war with the idea of teaching a lesson is practical and pragmatic, even in 2030.[3] However, this needs to be weighed against another school of thought that feels that China is concerned about its image as a responsible world power and would not risk a stalemate in a Sino-Indian conflict. In such an eventuality, China's political aim would be to capture its claimed areas.

Unlike China, whose four modernisations have made the country a global economic and defence player, the plausible scenario for Pakistani in 2030 is difficult to predict. Four broad visualisations denoted as the 4F scenarios emerge – functional, fragile, failing and fragmented Pakistan. Certainly, in the global interest, it would be ideal to have a functional Pakistan. In such a scenario, the Pakistani security forces come down heavily on radical elements. The Taliban are moderated and safe havens are denied to the Al Qaeda. Afghanistan is allowed to stabilise. The civil sector starts functioning and social indictors improve as a result of balance in economic policies. There is a general improvement in all spheres of public life but border issues with India remain unresolved despite good trade between the two countries.[4] In such a scenario, China would continue to be Pakistan's mentor and both could possibly collude in launching operations against India simultaneously. It would possibly be a very difficult proposition for India to manage a two-front war, but if the enemy wants to teach us a lesson, this would strategically be the best way to plan for facing both the Chinese and Pakistani preparations. Not only have India's armed forces to deal with two fronts but also with the insurgency problem which takes on another half front. The government and the security forces must strain every sinew and take positive steps to deal with the enemy on two fronts. Historically, Vietnam has simultaneously fought on two fronts successfully in 1979. Its forces captured Cambodia by defeating the forces of Pol Pot in the southwestern front, simultaneously defending the country against a Chinese offensive on the northern front. The result of this two-front war was that Cambodia was captured by the Vietnamese and the Chinese forces withdrew from all captured territories, having learnt the need to modernise and change their doctrines. It was the brilliance of Gen Von Nguyen Giap to have orchestrated this two-front war with military precision.

India has to be prepared for the worst contingency which would entail undertaking a two-front war with China in collusion with Pakistan against a nuclear backdrop. Our firepower resources have to be capable of dealing with the visualised threat in a network-centric environment. Accordingly, capability building must be accorded top priority. A report entitled "Non Alignment - 2.0: A Foreign and Strategic Policy for the 21$^{st}$ Century" was unveiled on March 2012 at a panel discussion in which the National Security Adviser Shri Shivshankar Menon, and his predecessors Shri MK Narayanan and Shri Brajesh Mishra, participated. The report contends that India cannot entirely dismiss the possibility of a major military offensive in Arunachal Pradesh or Ladakh. China will, for the foreseeable future, remain a significant foreign policy and security challenge for India. It is a country which impinges directly on India's geo-political space. As its economic and military capabilities expand, its power differential with India is likely to widen. The report goes on to say that in case of a military offensive or a territorial grab, India will need a mix of defensive and offensive capabilities to restore the *status quo ante*. The report recommends that in response to land grabs by China India should undertake similar actions across the LAC, a strategy of *quid pro quo*. Further, India must prepare itself to trigger an insurgency in the areas occupied by Chinese forces and develop firepower capabilities to interdict the logistics and military infrastructure in Tibet. Firepower must be developed by the Navy and Air Force to enhance our capabilities in the Indian Ocean.[5]

With a two-front war likely to be fought in 2030, asymmetries of firepower will make a major difference in offensive and defensive actions against our adversaries. To prepare for such an eventuality, our focus should be China. Our capability to handle China will result in handling Pakistan *en-passant*.

## Areas of Dispute

Our country is the world's seventh largest country in terms of area and is fortunate to have mountains, jungles, deserts and plains as part of its physical configuration. We have a border of 3,440 km with China and 3,310 km with Pakistan of which 790 km is the LoC. The LAC is presently the border with China but is not accepted as such by either side. The Chinese claim the entire state of Arunachal and their focus is on Tawang. The Indian side claims Aksai Chin, which was captured by the Chinese in 1962, and the Shaksgam Valley, which was ceded by Pakistan to China as a gesture of friendship in 1963. Both India and Pakistan lay claims to the state of J&K as also the Siachen Glacier. Further, there is a minor dispute on Sir Creek which is a 96-km strip of water between India and Pakistan. The creek divides the Kutch region of the Indian state of Gujarat from the Sindh province of Pakistan. The creek itself is located in uninhabited marshlands. During the monsoon season, between June and September, the creek floods its banks and envelops the low lying salty mudflats around it. The dispute lies in the interpretation of the maritime boundary which has to be resolved bilaterally.

Border disputes take a long time to be resolved and often become causes for conflicts. In a two-front war, all these areas would gain tremendous importance in the planning and execution of operations. Apart from these fronts, our relations with Sri Lanka, Bangladesh and Myanmar, though friendly, need careful nurturing. Both China and Pakistan are in commercial deals with these countries and are on the lookout for opportunities to improve defence relationships with them. Naturally, friendships, particularly in the defence field, are a cause of concern to India. Recently, India has been allotted two oil bocks in the South China Sea by Vietnam. China has objected seriously to this and its could lead to a faceoff. As China views the entire South China Sea as part of its Exclusive Economic Zone (EEZ), our armed forces must be prepared for such contingencies. Accordingly, our area of interest lies between the Strait of Hormuz and the South China Sea.

## Relative Strengths

China has the largest armed forces in the world. With a declared defence budget of US $ 106.4 billion, its capabilities have enhanced exponentially in all spheres. China compares its defence preparedness with that of the USA which is its sole strategic competitor. Its preparedness is directed with the aim of matching US capability in a full spectrum conflict. In viewing this, it is particularly aware of its drawbacks in firepower, particularly with regards to naval, amphibious, air and space platforms. Its focus is on integrating technology and using the asymmetry to its advantage by developing specialised missiles like the anti-aircraft carrier missile for engaging aircraft carrier group task forces, as also special forces, stealth fighters, SLBMs, competent information warriors, satellites providing navigation systems and, finally, a space station enabling research in the fields of astronomy, surveillance systems and, possibly, space-based weapons. China's present and future leaders all comprise technocrats who are leaving no stone unturned to bridge the technology gap with the USA and this would automatically put them way ahead of India.

At the national level, security is an uncommon and unknown word. Most of the people have little knowledge of strategy and the role the armed forces play in nation building and their role in developing Comprehensive National Power (CNP). The current Indian defence budget is about $ 37.8 billion which is about 36 percent of China's defence budget. The key equipment held by the Chinese comprises 7,400 battle tanks, 18,000 towed artillery guns, 1,200 self-propelled artillery guns, 7,500 air defence artillery guns, 1,669 fighter/ground attack aircraft, 71 submarines and 66 land-based ICBM launchers. Considering a two-front war, Pakistan has a defence budget of $ 6.41 billion and has 2,640 battle tanks, about 2,700 artillery guns, 1,414 aircraft, five submarines and combination of ballistic and cruise missiles. In comparison, India has 3,233 battle tanks, about 4,500 artillery guns, 800 fighter/ ground attack aircraft, 15 submarines, and is in the process of developing ICBMs. Numerically, India would be at a disadvantage in a two-front war as our assets are relatively less. Accordingly, we must do our best to avoid a two-front war, by statecraft and diplomacy, to avoid a difficult strategic situation. However, in case it is unavoidable, we have to enhance our force levels and procure additional equipment. To examine the relative firepower strengths of China, India and Pakistan, it is essential to know the firepower matrix of the PLA, the Pakistani armed forces and the Indian armed forces. The terrain, combined with the type of firepower, would determine the relative capabilities of these countries which would give us a clear picture once translated into their likely objectives.

The present Chinese military structure was created in the 1980s by Deng Xiaoping. The Chinese armed forces primarily comprise the PLA, the People's Armed Police Force (PAPF) and the militia. The PLA comprises the Army, Navy, Air Force and the Second Artillery. The PLA has seven Military Area Commands which are located at Shenyang, Beijing, Lanzhou, Jinan, Nanjing, Guangzhou and Chengdu. The Navy and Air Force are officially designated as the PLA Navy (PLAN) and PLA Air Force (PLA AF) respectively. The PAPF is responsible for maintenance of internal security, law and order, and to keep the country free from social disturbance. The Central Military Commission (CMC) exercises command and control over the armed forces and is directly under the Chinese Communist Party (CCP). The Chairman of the CMC is the commander of the armed forces. The Chinese Ministry of Defence does not exercise direct control over the PLA and is entrusted with the responsibility of the defence industry, construction of military facilities and foreign relationships.[6]

The ground forces of the PLA are divided into two constituents: the main force and the local force. The main forces are directly controlled by the PLA Headquarters and the local forces are under the control of the Military Region (MR). There are 18 Group Armies which are equivalent to our corps and these are the strike elements of the PLA, capable of being used for offensive operations. There are three types of Group Armies: Type 1 is for the plains, Type 2 is for the mountains and the jungles and Type 3 is for the frontier/coastal areas. For operations in the Tibetan plateau, Group Army Type 2 would be suitable. Type 2 (Mountain) Group Army is likely to have two to three infantry divisions, one artillery division / brigade and one anti-aircraft artillery brigade. The local forces are deployed in a specific geographical area in defence of land or coastal frontiers. In addition, they also share the responsibility for internal security.

Broadly, the local forces comprise the Border Defence Forces, Internal Defence Forces and Garrison Units. The Border Defence Forces are for early warning and reconnaissance, the Internal Defence Forces are for insurgency-prone areas like Tibet and Xinjiang, and Garrison Units are deployed in islands belonging to the People's Republic of China (PRC). They are equipped at a lower scale than the main force but gradually this is losing its distinction. China has Rapid Reaction Forces which comprise three Group Armies and six divisions. These are the 38 Group Army located at Baoding, 39 Group Army at Xiamen and the 54 Group Army at Xianning. In addition, there are six divisions from other Group Armies which have been equipped as Rapid Reaction Forces. 15 Airborne Corps, which is located at Wuhan, has three airborne divisions, which are capable of being transported by IL-76 aircraft. However, the total airlift is limited to one division at a time. The PLA ground forces have adequate firepower which emanates from 18 Group Armies and divisions as also brigades emanating from the local forces. Reportedly, there is a total of 61 infantry divisions, 10 armoured divisions, 3 airborne divisions, 7 artillery divisions, 12 independent armoured brigades, 29 independent infantry brigades, 18 independent artillery brigades, 17 anti-aircraft brigades and 9 Surface-to-Air-Missile (SAM) brigades. Firepower generated by the PLA ground forces is adequate to meet their operational requirements and they are constantly updating the same. There is a constant need of monitoring the forces in the Chengdu and Lanzhou Military Regions as they affect India directly for an offensive to be launched from Tibet.

The PLA Navy is aiming for blue water capability which would possibly take place by 2030. While it may not attain the global force level of the US Navy, it would be able to pose asymmetric threats like development of steep dive anti-aircraft missiles which would pose threats to carrier task forces. The PLAN has three fleets. The North Sea fleet headquarters is at Qingdao and it primarily undertakes offshore active defence of the Yellow China Sea; the East Sea Fleet has headquarters at Dongqain Lake (Ninsbo) with its focus on the East China Sea; and, finally, the South Sea Fleet which is headquartered at Zhangjiang, has a crucial task in furthering China's claims in the South China Sea. Its ships comprise one aircraft carrier (presently undergoing sea trials), three nuclear powered ballistic missile submarines (SSBNs), two Jia class and one Xia class, capable of firing SLBMs. Further, China has seven nuclear powered submarines, 76 other submarines, 27 destroyers, 35 frigates, 19 Landing Ships Transport (LST), 346 patrol vessels and numerous minesweepers and auxiliary ship. The Naval Aviation has 22 Sukhoi-30 Mark IIs, 12 Sukhoi-3s shipborne (under order), 165 fighter bombers which are capable of striking ships, 715 fighters capable of ground attack strikes, 24 maritime reconnaissance aircraft with capabilities for Anti-

Submarine Warfare (ASW) and helicopters as also transport aircraft. By 2030. China will have possibly three aircraft carriers and a modernised Navy capable of patrolling the Indian Ocean, the seas bordering the country and gradually making it to the fringes of the Pacific Ocean. Chinese ships using ports in Sri Lanka, Pakistan and Myanmar would be able to challenge India's Exclusive Economic Zone (EEZ). Further, our oil blocks in the South China Sea would need protection against a possible Chinese offensive. This would need to be dovetailed in our preparations.

The PLAAF is the third largest Air Force in the world and has genuinely started modernising after the first Gulf War when China realised that there was a need to possess air power despite being a nuclear power. The shock and awe created by the US Air Force woke up the PLAAF and thereafter, it has been on a furious modernisation drive. It was at this time that the Soviet Union broke up and the newly formed Russian government was in serious difficulties. With its economy in good shape, China signed a major deal for the purchase of 24 SU-27s (roughly in the US F-15 class), with a provision for the local manufacture of another 200 aircraft. China also purchased large numbers of the Rolls Royce Spey–200 engines from the UK for the locally developed JH-7/ FB-7 fighter bombers. Further, it accessed technology from Israel and also purchased the Russian aero-engines, the RD-93 and AI-31 F, to kick-start two other local fighter programmes. The result was the JF-17 Thunder which was jointly produced with Pakistan, and the J-10, based on the Israeli designer fighter, the Lavi, of the F-16 class. Israel had abandoned the Lavi programme under pressure from the US. Pakistan, incidentally, had supplied a crashed F-16 and a dud air-launched Tomahawk cruise missile to China for reverse engineering. The Chinese are attempting to copy the AI-31 turbo fan engine that powers the SU-30 and this should be completed shortly. They have locally produced the WZ-10 as an attack helicopter. Further, they have modified the respected AN-12 with powerful engines and advanced avionics into an Airborne Early Warning and Control (AEW&C) aircraft. The H-6 (the Chinese version of the 1950 vintage TU-16 bomber) has been modified with the more powerful Russian D-30 KP engine and is still operational. These aircraft are used for flight refuelling, electronic surveillance and to deliver anti-ship cruise missiles from safe standoff distances. The Chinese have also test flown the J-20 , their fifth generation stealth fighter, and it is anticipated it would be inducted possibly in another five years.[7]

According to the *Military Balance 2011,* the PLAAF has about 1,687 combat aircraft. The Chinese are capable of manufacturing 40 to 50 combat aircraft every year. They also have about 550 transport aircraft. The aircraft are grouped into 45 divisions each having three regiments. Each regiment has three squadrons and each squadron has three flights, with each flight having three to four aircraft. With the success of UAV technology in recent global conflicts, China has inducted numerous UAVs and UCAVs in the PLA. Indigenous development programmes and covert tie-ups with Yamaha of Japan and Israel Aerospace Industries (IAI), Malat, of Israel, have resulted in the development and acquisition of numerous UAVs and UCAVs. The UAVs are the ASN 105, 229A, Japanese RMAX, Shenyang BA-5 and WZ -5. The UCAVs are the Israeli Harpy, Pterodactyl, Pterosaur, WJ-600 and CH-3. The UAVs and UCAVs are capable of surveillance, engagement of targets and PSDA. These unmanned systems are force multipliers and China is integrating the UAVs intelligently to enhance its firepower capability. The PLAAF has numerous airfields, including five in Tibet. These are Gongar, Pangta, Linchi, Hoping and Gar Gunsa.

The PLAAF has no combat experience nor has it participated in exercises with air forces other than a few aircraft in Turkey in 2011. Its equipment is, by and large, of the second and third generation, and it is indigenising as also modernising at an accelerated pace. Further, the type and variety of weapons, especially cruise missiles, UAVs and UCAVs as also the focus on space-based systems such as Glonass/GPS reconnaissance satellites indicate that the PLAAF is at par with other Air Forces in grasping details of employment of air power.

The Second Artillery Corps is the strategic missile force of the People's Republic of China. The Second Artillery was established on July 01, 1966, and made its first public appearance on October 01, 1984, The operational headquarters is located at Qinghe. The Second Artillery Corps is under the direct command of the Chinese Central Military Commission. The Second Artillery operates both conventional and nuclear missiles. Further, China operates both ballistic and cruise missiles. Due to limited information, the number of nuclear warheads held by China is not accurately known. It is estimated, as per reports, that it has about 400 active warheads and about 3,000 warheads hidden within an extensive tunnel system referred to as an underground great wall.[8]

China possesses one of the largest land-based missile forces in the world. The Second Artillery Corps is capable of inundating the region surrounding China with myriad quantities of ballistic and conventional missiles. China fired its first ICBM in 1980, the DF-5, capable of engaging targets in both the US and Soviet Union. In 1981, China launched three satellites from a single vehicle, thereby attaining the capability of launching MIRVs. By 1986, China developed a credible deterrent force with land, sea and air elements. Missiles have become the mainstay of the Second Artillery. The Chinese land-based ballistic and cruise missile force comprises 38 operational missile units. There are eight units supporting ICBMs and the remaining are mobile theatre-based systems. The units are organised on what the PLA terms as bases. There are six bases located in different geographic locations. Each base has numerous subordinate missile brigades, with each brigade maintaining one or more garrison, various underground facilities, rail transfer points and field launch positions. It is indeed difficult to identify the field launch pads. The process used for identification is a hard concrete pad where the associated missile will be erected for launch. It is observed that the typical dimensions for the DF-11 is 15 m in length, for the DF-15, it is 26 m in length and for the DF-21, 45 m in length.

The location of the six bases has been done keeping the strategic aspects under consideration. The 51$^{st}$ base consists of six missile battalions in northeastern China, facing southeast Russia, South Korea and Japan. The 52$^{nd}$ base consists of 12 missile battalions in southeast China, facing Taiwan. The 53$^{rd}$ base consists of four missile battalions in southern China, facing Vietnam, and possibly Tibet, at extreme ranges. The 54$^{th}$ base has six missile battalions in eastern China catering for Taiwan, southern Japan and the southern portions of the Republic of Korea. The 55$^{th}$ base has two missile battalions in southern China targeting Tibet and Vietnam. The 56$^{th}$ base is formed of eight missile battalions in northern China, mainly catering for targets in Russia. These units are extremely flexible and, except for limited numbers in silos, can be moved to different locations by road or rail.

Currently, China's missile inventory comprises a variety of ballistic and cruise missiles. Essential missiles are the DF-2 (CSS-2) with a range of 2,650 km, DF-4 (CSS-3) with a range of 2,200 km, the modified DF-5A (CSS-4, Mod-2), an ICBM with MIRV attaining a maximum range of 13,000 km, DF-21A (CSS-5, Mod-2) with a range of 2,150 km, DF-15 C (CSS-6, Mod 3, modified M9) attaining a range of 750 km, DF-11A (CSS-7, Mod-2, modified M-11), maximum range of 500 km, JL-2 (CSS NX-14) attaining a range of 7,200 km, DF-31 (CSS- 10, Mod-1) with a range of 8,000 km and, finally, DF-31A (CSS-10, Mod-2) with a range of 11,200 km and MIRV capability. On July 24, 2012, China tested its new DF-41 ICBM with a range of 15,000 km that can cover the entire USA. China has modified the DF-21D with possibly a steep dive capability for engagement of ships. China has numerous cruise missiles which are the SY-1, HY, FL,

YJ. C 701, C 801 and these have a range of between 300 to 600 km.

There are about 650 DF-11A (M-11) and DF-15 C (M-11) missiles deployed opposite Taiwan, and several dozens of DF-4 and DF-21A medium range missiles that can target India, Russia and Japan. There are about 18 to 24 DF-5As which can reach locations in Europe and the United States. China is in the process of deploying the newly developed DF-31A and the JL-2 against any worthwhile target. The Chinese Second Artillery has the flexibility of changing the quantum of missiles with ease and confidence. Further, China has moved to solid propellants with respect to the DF-11A, DF-15A, DF-21A, DF-31 A, DF-41 and JL-2.

China would like to use its missiles effectively on Indian targets in a full spectrum conflict. Hans Kristensen, the Director of the Nuclear Information Project, Federation of American Security (FAS), has posted details of deployment of nuclear missiles in central China and Tibet. He has accurately tracked deployment of Medium Range Ballistic Missiles (MRBMs) in Tsonub Mongolia, Qinghai and Tibet. Google photographs depict about 60 launch pads near the city of Delingha and the Da Qaida basin. Further, there is infrastructural development for command and control facilities in the city of Delingha. The city lies on the northeastern edge of the Da Qaida basin and is located 500 km west of Xining. There are reportedly three brigades of DF-21A missiles located out here. Each brigade comprises three to four battalions and each battalion has three to four companies each with a launcher. Thus, there is a total of 27 to 36 launchers of the DF-21A with a range of 2,150 km in this region. Delingha is located about 2,000 km from Delhi and this could be easily engaged with the available missile system. The railway network has reached Lhasa in 2006 and the DF-21, mounted on a Transport Erector Launcher (TEL), can be deployed south of Lhasa to engage Chennai. The upgraded road network also enables easy movement of the missile to Niyachi which is 400 km southeast of Lhasa and north of Arunachal Pradesh. To augment the resources, there are missile units in Kunming which can engage targets in India at maximum ranges. Further, China's ICBMs with MIRVs can cover our entire country with ease.

China has, since the Korean War, focussed on explorations in outer space. Effective firepower would need accurate intelligence based on satellite inputs, which would result in correct selection of targets, their effective engagement and PSDA. $C^4I^2SR$ is a major component for command and control of guns, rockets, missiles, aircraft, ships and submarines. The entire system depends on effective inputs of satellites deployed in outer space. China has a large number of satellites providing military surveillance, communications and weather inputs. These would ensure effective engagement of targets in Arunachal, Sikkim, the UP-Tibet border, eastern Ladakh, ships in the Indian Ocean and oil blocks in the South China Sea. China is improving its overall capabilities in outer space. Space stations on the lines of the International Space Station are going to be established by 2020. The target vehicle Tiangong 1was established on September 29, 2011, which can host a crew of three for a limited period. The Tiangong 2 is to be launched in 2013 and this would be a space laboratory with the capability of accommodating a crew of three for 20 days. This will be followed by Tiangong 3 which would be launched in 2015 with the ability to provide 40 days life support for a crew of three. The complete system would be operationalised by 2020, enabling human control of outer space by Chinese space specialists.

Navigation is a major factor in providing accurate firepower. To avoid dependence on GPS or Glonass, China has developed, and is gradually making, its satellite navigation system operational. In October 2000, China launched the BeiDou-1A satellite followed by the BeiDou-1B on December 21, 2000 and BeiDou-1C on May 25, 2003, thus, providing a navigation system comprising three satellites giving locations over China to an accuracy of 10 m. Over

1,000 BeiDou-1 terminals were used in the Sichuan earthquake in 2008 for providing information from the earthquake site and, as of date, border guards are equipped with BeiDou-1 devices. The global system (BeiDou-2) is a new system which in its final shape would have 35 satellites. These would be sub-divided into five geo-stationary orbit satellites for backward compatibility with BeiDou-1 and 30 non-geo-stationary satellites (27in medium earth orbit and 3 in inclined geo-stationary orbit) that will have global coverage. Further, free service with 10 m accuracy would be provided for civilian use. The licensed service will be more accurate and would be for military purposes. The first satellite was launched on April 13, 2007, and 10 satellites would be launched by 2012 and would offer services for the Asia-Pacific region. The entire system would be operational by 2020.[9] China would be self-reliant in navigation, thereby giving the nation comprehensive capability to launch its weapon systems at any point on the globe.

China also successfully fired an anti-satellite weapon on January 17, 2007 and is currently docking satellites in space. This gives the Chinese the capability to interfere with the transmission signals of other satellites, thereby attenuating their capabilities. It is also reported that China is developing DEWs to be operated in outer space. Use of cyber war with kinetic or thermal direct energy weapons would give China the capability to destroy satellites, missiles in their boost phase as also aircraft, surface ships and possibly land targets.

The Pakistan Army is a volunteer professional fighting force and as per the International Institute for Strategic Studies (IISS), it has an active strength of 550,000. The Army is planning to raise three regional Command Headquarters as the Northern, Central and Southern Commands. Locations of these headquarters are under finalisation. Pakistan has nine corps, a Force Command Northern Area (FCNA), 19 infantry divisions, one artillery division and two armoured divisions, along with six independent mechanised brigades, seven independent armoured brigades, nine independent artillery brigades and seven engineer brigades. The Army also has an Air Defence Command with three Air Defence (AD) groups and eight AD brigades.

In terms of firepower, the deployment of formations is balanced and starts with the FCNA at Gilgit with two divisions. This is opposite Ladakh and the Kashmir Valley. South of FCNA we have X Corps, with three divisions opposite the remaining portion of Jammu and Kashmir, XXX Corps at Gujranwala with two infantry divisions opposite Jammu city and northern Punjab. South of this formation, there is IV Corps at Lahore with three infantry divisions, opposite Punjab , XXXI Corps at Bhawalpur with two infantry divisions and, finally, V Corps at Karachi with two infantry divisions. Pakistan has reserves for the plains comprising Army Reserve North consisting of I Corps with one armoured division, two infantry divisions located at Mangla and II Corps as Army Reserve South with one armoured division, one infantry division located at Multan. Pakistan also has XI Corps with two infantry divisions located at Peshawar, presently being employed at the Federally Administered Tribal Areas (FATA) and XII Corps with two infantry divisions located at Quetta. Both these corps have the flexibility of operating in the plains or mountains based on the overall strategic situation. Firepower elements exist in all these formations. There is an artillery brigade with each division and an independent artillery brigade at the corps level. Pakistan also has surveillance and target acquisition capability at the artillery brigade level and has raised an artillery division comprising the corps component of two artillery brigades and an air defence unit.

Weapons with the land forces of Pakistan are a combination of modern and obsolescent equipment. As far as the artillery is concerned, Pakistan has self-propelled artillery comprising limited numbers of the 203 mm (old 8 inch), 155 mm M 109 howitzer and about 90 Norinco 155 mm SH1 6x6 wheeled chassis. Towed guns comprise some M 115, 203 mm, twelve MKEK 155mm Panter, 124 M198, 155 mm, 65 M114, 155 mm, 300 M 109

A2 155mm, 115 M 109 A5 155 mm, 200 Type 59, 130 mm, 600 Type 60, 122mm, 300 M101 105 mm, 200 Type 56, 85mm and 120 mm mortars. Pakistan is also in possession of about 40 BM 21, 122 mm multiple launcher rockets and unknown, limited numbers of A-100, 300 mm rockets. Apart from this, Pakistan has a mechanised force of about 2,400 tanks and about 1,600 Armoured Personnel Carriers (APCs). Broadly, there are about 350 Al Khalids, 320 T-80UDs,250 Type 69s, 500 Al Zarars , about 100 T-60/63 light tanks and the remaining T-54/55 tanks. As far as the APCs are concerned, there are about 800 M-113s, 149 Al Fahds, approximately 500 Talhas, 120 BTR-70s and Hamzas under procurement. Pakistan has an air defence system comprising the ZU-23, 30mm, 35 mm, 40 mm, 57mm guns and 400 RBS-70, 6 CS-AIs (SAM-2), 100 or more Stinger missiles. It also has the medium altitude air defence system and the high altitude air defence system comprising the Spada-2000, Bofors RBS 23 in the former and HQ-2B SAM as also HQ-9 respectively. Pakistan has an Army Aviation component of 11 helicopter squadrons. Out of these, two squadrons at Multan, equipped with AH-1S and AH-1F Cobras, are attack helicopters. The Pakistan Army is also inducting UAVs for surveillance.[10]

The Pakistan Navy is a small force of 25,000 and has five naval commands which are the Fleet, Karachi Shore Establishments, Coast Shore Establishments, Logistics and Northern. The main constituents of firepower in the Pakistan Navy are frigates, submarines, light craft aviation and a battalion of marines. As far as frigates are concerned, there are four Zulfiqar class (F-22 type), six Tariq class (ex UK Type 21) and four Alamgir class (ex US FFG-7). Submarines held by the Pakistan Navy are three Khalid class (Agosta 90B), two Hashmat class (Agosta 70) and three SX 404 class (midgets). Pakistan is presumably developing a nuclear submarine which is expected to be developed earliest by 2020. All the submarines are equipped with anti-ship missiles which can be fired while submerged. The Khalid class can fire the Exocet missiles whereas the Hashmat class can fire the Harpoon missiles. For the last 10 years, Pakistan has been experimenting with the Babur Land Attack Cruise Missile (LACM) which has the capability of firing both conventional and nuclear warheads. Corvettes and missile boats operated by the Pakistan Navy comprise four Jalat class missile boats armed with four C-802/C-803 Chinese anti-ship missiles. There are also two ships of the Quwwat class and one ship of the Larkana class amongst the light craft.

The Naval Aviation is a potent arm of the Pakistan Navy. Since 1993-94, naval fighter pilots have been flying the mirage V Rose fighter jets, equipped with Exocet missiles. Presumably, there are 32 Mirage Vs located at Masroor air base in Karachi. The Pakistan Naval Aviation consists of a variety of helicopters and aircraft. There are three Maritime Patrol (MP) / Anti-Submarine Warfare (ASW) squadrons. One squadron has two Atlantiques which are reconnaissance, surveillance and ASW aircraft, the second squadron has six F-27s and the third squadron has 10 P3C Orion naval surveillance craft, ASW, airborne early warning and bombers. Apart from these, there are numerous anti-ship and ASW helicopters. It is reported that the Pakistan Navy is possibly inducting UAVs which could be either Nexcom, Burraq or Satuma Spy. Thus, the Navy is modernising and by 2030, will have an additional port at Gwadar and play an important role in the Arabian Sea.[11]

The Pakistan Air Force (PAF), with its firepower assets, provides air defence to Pakistan in consonance with the Army and Navy. The PAF operates over 500 fighter jets along with some reconnaissance, transport and trainer aircraft. Under the Armed Forces Development Programme 2019, initiated by the government, the PAF is set to undergo modernisation by acquiring new aircraft, mid-air refuellers, UAVs and long range missiles. The major air bases are at Karachi, Peshawar, Chaklala, Quetta, Kamra, Mianwali, Shorkot, Jacobabad and Sargodha.[12]

The PAF has three commands: Southern Air Command located at Karachi, Central Air Command at Sargodha and Northern Air Command at Peshawar. Currently, the PAF operates four different types of aircraft and there are about 20 frontline squadrons. The PAF is primarily geared for air defence operations. The primary air defence fighter is the Chengdu F-7 of which two variants are in service. These are the F-7P, of which there are around 120, and 60 F-7PGs. An upgraded variant, the F-7M, incorporates many specific PAF modifications such as two extra weapon stations, an extra 30 mm cannon and an Italian designed Grifo-7 multi-mode radar. The F-7 is also used to perform limited strike duties. The second fighter aircraft is the French designed Mirage-3 and Mirage-5. The difference between the two aircraft is the shape of the nose and avionics. Mirage-3 fighters are geared towards fighting multiple missions, including interception and strike, whereas Mirage-5 fighters are more focussed on strike missions. There are about 150 aircraft and all of them have undergone upgradation. The JF-17 is a fighter jointly developed with China that is currently being inducted by the PAF and is expected to gradually replace all Mirage-3/5 and F-7 by 2015. In all, 250-300 aircraft are planned to be built, with improved airframe, avionics and engines. Presently, two squadrons are being equipped and the first Pakistani built JF-17 manufactured at the Pakistan Aeronautical Complex was rolled out and handed over to the PAF on November 23, 2009. The most capable fighter in the PAF is the F-16. Presently, there are about 22 F-16As, 20 F-16Bs, 12 F-16Cs and a few F-16Ds. Two squadrons of an advanced variant of the Chinese J-10 known as the FC-20 are to be inducted into the PAF by 2015. It is anticipated that a total of 150 fighters is planned to be inducted in the near future.

The PAF has seven Saab 2000 Erieye aircraft which are meant to perform Airborne Early Warning and Control (AEW&C). Each aircraft is fitted with five operator stations and four command stations. The aircraft's Erieye radar has a range of 450 km and is also capable of identifying the type of aircraft and the weapons it is carrying. The Erieye is connected via data link to the PAF's command and control ground environment as well as combat aircraft such as the F-16. Reports indicate that four Chinese ZDK-03 AEW&C aircraft have also been ordered. These are believed to be the PAF specific version of the KJ-200 incorporating the Chinese AESA radar. These would be able to direct aircraft against air and ground targets, thereby optimising their firepower assets.

The PAF has a veritable SAM system which is deployed to defend critical Vulnerable Areas (VAs) and Vulnerable Points (VPs). A short to medium range air defence system in use is the Crotale. The PAF had inducted 11 Crotale 2000 acquisition units and 23 missile batteries in 1976. They were upgraded to Crotale 4000 standard, thereby increasing the range from 20 to 30 km. The SPADA 2000 system is likely to replace the Crotale, thereby giving the capability to enhance the range by another 10 km. Pakistan has also acquired the RBS-70, a low altitude air defence system that fires laser beam riding missiles. This is also known as the VSHORAD (Very Short Range Air Defence) missile system which in its latest upgrade is unjammable and has a maximum range of 8 km. Pakistan has three more man-portable systems which are the Anza Mk-3, manufactured by Kahuta Research Laboratories, the Mistral low altitude man-portable air defence system of French origin, and the FIM-92 Stinger, a low altitude man-portable air defence system of US origin.[13]

The PAF is gradually inducting UAVs and UCAVs in its weapons inventory. Presently, it has 25 Selex Galileo Falco UAVs. These are being used for surveillance. About 45 BRAVO (SATUMA JASOOS II) tactical reconnaissance UAVs were formally inducted in 2009 and are being primarily being utilised for training. Pakistan is trying to develop the Burraq which is a reconnaissance strike UCAV and is in an early stage of development.

Pakistan formed the National Command Authority (NCA) on February 02, 2000, to oversee the country's strategic nuclear arsenal and related organisations. The NCA operates with a unified central command structure and comprises the Strategic Plans Division (SPD) and Strategic Forces Commands of the three Services, the Army, Navy and Air Force. The SPD has an elaborate security division which includes a counter-intelligence network to safeguard the activities of the strategic organisations. The missiles held by Pakistan are the Hatf-I/-IA with a range of 80 to 100 km, Hatf-II with a range of 300 km (Chinese M-11), Hatf-III with 600-800 km, Ghauri-I with 1,500 km, Ghauri-II with 1,500 to 2,300 km, Shaheen-I ranging 750 km (Chinese M-9), Shaheen-II with 2,500 km. The Hatf and Shaheen missiles use solid propellant; the Ghauri is based on liquid propellant. On April 19, 2011, Pakistan test-fired a solid-fuelled battlefield ranged ballistic missile which can be fired from multi tubes at a range of 60 km. It had a TEL as launch platform and a sub-kiloton warhead. This is to counter India's Cold Start doctrine. It is reported that Pakistan would like to use this tactical nuclear weapon on the Indian assaulting forces in their launch pads and assembly areas prior to reaching the objective. This would imply that Pakistan would like to use nuclear forces on Indian mechanised spearheads, prior to establishment of contact.

India is a democratic country which has to be prepared to fight a two-front war. It has hostile borders with both China and Pakistan. The Indian Army's strength is about 1.1 million combatants who are being regularly trained to fight a two-front war. There are six Operational Commands and a Training Command. There are six Field Armies, each capable of employing two to three corps. Firepower constituents are present in each Field Army. Presently, the Indian Army has 13 corps out of which three are strike corps designated for offensive operations in the plains. Each corps has two to three divisions and in total, there are three armoured divisions, four RAPIDs (Reorganised Army Plains Infantry Division), 18 infantry divisions, 12 mountain divisions as also three artillery divisions. In addition, there are six independent armoured brigades, one independent parachute brigade, two air defence groups (self-propelled), six air defence brigades, as also four engineer brigades. Each division has its own artillery brigade, and at the corps level, there is an artillery brigade except in the strike corps which has an artillery division. There are plans to raise two additional divisions and a strike corps for the mountains in the 12$^{th}$ Plan (2017-22).

The main equipment held by the Indian Army to provide firepower in terms of surveillance, reconnaissance, target acquisition and engagement is with the fighting formations. Essentially, as far as tanks are concerned, there are 124 Arjuns, 657 T-90s and 1,700 T-72s. There are about 1,600 BMP1/BMP2 as also 100 CASSPIR Mk2 mine protecting vehicles. With respect to artillery, there are about 2,500 105 mm Indian field guns/light field guns, 300 130 mm M-46 medium guns, 180 modified 155 mm M-46 guns, 410-155 mm Bofors guns, 130 Grad BM-21 rockets , 40-214 mm Pinaka multiple launcher rocket systems and 60-300 mm Smerch rocket systems. The artillery has numerous kinds of surveillance and target acquisition equipment, comprising about 40 Heron UAVs, 12 Searcher Mark-2 UAVs, 8 Searcher Mark-1 UAVs, 12 weapon locating radars, about 100 ELM-2130 battlefield surveillance radars and about 350 LORROS. The field artillery modernisation plan calls for all the guns to be standardised to a 155 mm calibre. Trials have been completed for the 155mm (39 calibre) ultra light howitzer, 155mm (52calibre) wheeled SP gun and 155 mm (39 calibre) OFB manufactured guns. Trials are likely to take place for the 155 mm (52 calibre) Tracked SP gun shortly. The Request For Proposal (RFP) has been issued for the 155 mm (52 calibre) towed gun and is under issue for the 155mm (52 calibre) Mounted Gun System (MGS). In terms of Anti-Tank Guided Weapons (ATGW), the Indian Army has the Milan and Konkurs ATGW. Negotiations are on for procuring either the Javelin from the USA or the Spike from Israel.

The Army Air Defence presently comprises equipment which needs to be modernised. These are in terms of anti-aircraft guns and anti-aircraft missiles. The anti-aircraft guns presently held are the L-40/70, 23mm twin barrelled guns, the 23mm quadruple guns mounted on the tracked vehicle Schilka, and Tunguska gun missile system mounted on tracks. The Tunguska system primarily comprises the 2S6 combat vehicle which uses the GM-352M chassis, radar system IRL144 (NATO: Hot Shot), dual 2A38M 30mm cannons with two 9M311-M1 missiles with a range of 10 km. The Surface-to-Air-Missiles (SAMs) comprise the SA-6, SA-7, SA-8, SA-9, SA-13, and SA-18. The Army Air Defence has placed orders for two Akash SAM regiments each constituting four batteries. Further efforts are on to procure a quick reaction SAM at the earliest. The Army has an Aviation Branch whose primary task is to provide surveillance, direction of artillery fire on depth targets and, whereever possible, act as an airborne Forward Air Controller (FAC) for fighter aircraft. Presently, there are about 60 Chetaks (Alouette II), 120 Cheetahs (Alouette III), 40 Dhruv Advanced Light Helicopters (ALH) manufactured indigenously. A weaponised version of the Dhruv is under development. There is a deliberate effort on to get about 197 helicopters in conjunction with the Indian Air Force to replace the current lot of Cheetahs and Chetaks. This is proposed to be done at the earliest and, hopefully, will be completed by 2017.

The other aspect which merits attention is the induction of the supersonic cruise missile BrahMos as a weapon of the Indian Army. The missile has a range of 290 km and has been inducted in the Indian Army. The missile is produced by a joint venture company whose partners are the DRDO, Government of India, and the Russian Federation. The missile is extremely accurate and is capable of precise engagements and surgical strikes. Recently, the missile has demonstrated its steep dive capability which enables the system to be deployed by the Indian Army on its northern and northeastern borders. The missile carries a conventional warhead facilitating its usage in conventional conflicts with a high degree of accuracy.[14]

Firepower forms a major constituent of the Indian Navy as practically all engagements are going to occur in the Beyond Visual Range (BVR). The Indian Navy has a strength of 55,000, out of whom 7,000 are a part of the naval aviation and 1,200 are Marines. Presently, the Navy has three commands: the Western Naval Command, Eastern Naval Command and Southern Naval Command. The Western and Eastern Naval Commands are operational whereas the Southern Naval Command looks after the training of the Indian Navy. Presently, the Navy has the Western Fleet and the Eastern Fleet. A document pertaining to naval force level structure states that the Navy would have two carrier task forces with a total of 125 warships. There would be a minimum of 24 submarines. Further, a special anti-terrorist force is being created with 1,000 personnel and 80 fast interceptor boats.[15]

The major warships with the Indian Navy are one aircraft carrier, the INS *Viraat* which is operational at present. Further, the INS *Vikramaditya* (ex *Admiral Goroshkov*) is expected to be delivered from Russia in 2013. It may be pertinent to note that an indigenous aircraft carrier of 37,500 tons is under construction at Kochi Shipyard, due for commissioning by 2015. In addition, plans have been formulated for two follow-on vessels, the IAC2 and IAC3. By 2030, possibly, there would be definitely two if not three, carrier task forces. They would provide veritable naval firepower in the Arabian Sea, Bay of Bengal and Indian Ocean. The four types of destroyers—the Delhi class-3, Rajput class-5, Ranvir class-2 and Kolkata class-4—making a total of 14, would definitely provide a battleworthy naval asset for sea control and sea denial. In conjunction with the destroyers, the Navy has six types of frigates. The Talwar class has four frigates, including the latest INS *Teg* which was commissioned on April 27, 2012, at the Yantar Shipyard in Russia. This is a modern stealth warship which is an advanced version of the Talwar class with a plethora of firepower systems. The weapons suite includes the surface-to-air and surface-to-surface BrahMos missile systems. In addition, it includes a 100 mm gun, torpedo tubes and anti-submarine rockets. There are orders for two more of this category to be constructed for the Indian Navy. The other stealth multi-role frigate comprises the Shivalik class (Project 17) which has two ships. Apart from these two types, the other versions are the Brahmaputra class (Project 16 A)-3 ships, Godavari class (Project 16)-3 ships and Nilgiri class-1 ship. There are 16 submarines presently in service. There are *Sindhughosh* (Kilo)-11, *Shishukumar* (Type 209)-4 and one Akula class nuclear attack submarine on lease from Russia with our Navy for a period of 10 years. India is also producing its own nuclear powered submarine, the INS *Arihant,* which would be shortly undergoing sea acceptance trials by the Indian Navy. It is projected to enter service by 2015 and a class of six submarines is planned in this category. Six Scorpene class submarines are on order for construction indigenously. The project is running behind schedule. The first submarine is likely to be commissioned by August 2015 and the sixth, possibly by May 2019. The RFP for a bigger size submarine has been circulated with two to be built in foreign yards, and four to be constructed locally.

The Indian Navy also has a lighter class of warships. Presently, there are 24 Corvettes in service. They comprise the *Kora* (guided-missile Corvettes)-4, *Khukri*-4, *Veer* (Tarantul I, missile boat)-12, *Abhay* (Pauk II, anti-submarine Corvette)-4. In addition, there are 10 amphibious warfare vessels which include the recently acquired INS *Jalashwa* from the USA, 29 patrol vessels and minesweepers.

The naval aviation has 7,000 personnel and a variety of firepower assets. There are two fighter squadrons which are shipborne. Eleven Sea Harriers are operational on the INS *Viraat* and the MiG-29K squadron is being formed for the INS *Vikramaditya.* Twelve MiG-29Ks and four MiG-19KUBs have been delivered. A follow-up order of 29 MiG-29Ks has been placed for the aircraft carrier being built at Kochi. Surveillance and reconnaissance which facilitate firepower are carried out by the aviation wing of the Indian Navy. There are four maritime reconnaissance squadrons, "Albatross" squadron with 8 Tu-142Ms at Arakkonam; "Winged Stallions" at Goa with 5 IL-38s; "Hawks" at Port Blair; and "Flying Fish" at Kochi, each with 12 Dornier-228s.In as much as helicopters are concerned, they play an important role in surveillance and Anti-Submarine Warfare (ASW). In this case, there are four squadrons, "Eagles" at Vishakapatnam with 6 Ka-25 Hormones, 5 Ka-28 Helixs; "Harpoons" at Mumbai with 20 Seakings; "Flaming Harrows" at Kochi with 12 Seakings, and "Falcons" at Mumbai with 6 Ka-28 Helixs. In conjunction, the Navy has one electronic warfare squadron at Goa, two UAV squadrons with 8 Searcher Mark-II and 6 Herons. The Navy is in the process of procuring 75 multi-role helicopters at an estimated cost of $ 4 billion. A global RFP is to be issued and presently trials are on for acquisition of 16 multi-role helicopters in which the Sikorsky S-70 Bravo and European NH-90 are being evaluated.[16]

Firepower assets are also available with the Coast Guard which is an independent service but could work in conjunction with the Indian Navy. The Coast Guard has 15 Offshore Patrol Vessels (OPVs) as also 12 Dornier 228s. Further, three OPVs, 4 Inshore Patrol Vessels (IPVs), 11 interceptor boats and an unspecified numbers of T-80 coastal fast attack craft (Israeli Super Dvora MkII) are being procured.

The Indian Air Force (IAF) has a strength of 170,000, and has five Operational Commands, one Training Command and one Maintenance Command. The location of the headquarters of these commands are Western-New Delhi, Southwestern-Gandhinagar, Eastern-Shillong, Central-Allahabad, Southern-Trivandrum, Training-Bangalore and Maintenance-Nagpur. For combat, the IAF has 12 air defence squadrons, 17 fighter/ground attack squadrons, one maritime strike squadron, three close air support helicopter units, nine transport squadrons, one tanker squadron and one AEW&C squadron.

The IAF is in the process of modernisation and has a wide range of aircraft capable of delivering firepower against air, land and sea targets. The fighter aircraft comprise the potent Sukhoi 30MK1, of which about 165 are held and 107 are being manufactured by HAL. This is a multi-role air superiority fighter whose firepower assets are of a very high magnitude. It carries all kinds of weaponry, from a gun, all types of missiles and laser guided bombs. There are 67 MiG-29 air superiority fighters which are being upgraded to the MiG-29 SMT standard and 49 Mirage-2000s which are in the process of upgradation. The IAF has also inducted the indigenous fighter Tejas. Seven of them are in service and 40 are indented on HAL. Apart from these, in the fighter category we have the MiG-21 Bis-55, MiG-21M/MF-80 and MiG-21-93 Bison-120. Fighters in the ground attack role are 128 Jaguars and 120 MiG-27s. The IAF is procuring 126 state-of-the-art Rafale Medium Range Multi-Role Combat Aircraft (MMRCA) from Dassault, France, for which price negotiations are on. The Rafale would replace the MiG-21 on induction. This would possibly begin by 2015. The IAF has also formulated its strategy to procure the Fifth Generation Fighter Aircraft (FGFA). The development would be jointly undertaken between Sukhoi of Russia and HAL of India. The project is derived from the PAK FA in which T-50 is the prototype. The Indian version will officially be designated as FGFA. As per the latest statements by the government, the aircraft is likely to be introduced in the IAF in 2019. The IAF plans to induct 214 of these aircraft. Out of these, 166 will be single-seater and 48 will have a crew of two pilots. There is no reliable information regarding the design, but Russian specialists have commented off the record that the aircraft will be stealthy, have the ability to super cruise and be fitted with the next generation of air-to-air, air-to-surface, air-to-ship missiles and incorporate the AESA radar. The FGFA will in all probability be fully inducted by 2030 to replace the current Sukoi 30 Mk1.

Firepower in the IAF is also undertaken by helicopters and transport aircraft. The IAF currently holds about 20 MI-35 Hinds and is in the process of procuring 20 AH-64D Apaches from the USA. The IAF has 45 Dhruv utility-cum-attack helicopters and has placed an order on HAL for an additional 145. HAL is also developing a light combat helicopter and the IAF has placed an order for 65. Three IL-76 Phalcons have been modified to AEW&C aircraft and the IL-78 is capable of in-flight refuelling, thereby, enhancing the range of our fighter aircraft. There are also three A-50EI which would be performing the task AEW&C, thereby, enhancing the capability of command and control from the air.

The IAF has other surveillance assets like UAVs and aerostats. The UAVs held by the IAF are the Searcher MK-II and Heron. While the Searcher MK-II operates up to an altitude of 15,000 ft, the Heron operates up to an altitude of 30,000 ft. They are fine-tuned for the mountains and are meeting the expectations of spot surveillance and confirmatory reconnaissance. Currently, the IAF has two aerostats, one deployed in Western Air Command and the other deployed in Southwestern Air Command. The IAF has inducted the Harop UCAV which has a range of 1,000 km and can loiter and destroy targets with pinpoint accuracy in depth. Endurance of the UCAV is six hours.

The Ministry of Defence, on the insistence of the three Services, formed an Integrated Space Cell on June 10, 2010, for optimisation of Indian space assets. The space cell will ultimately be built to a Space Force Command. With all the powers making forays into outer space , the organisation will have to plan to defend existing space assets as also develop plans for deployment of firepower assets in outer space in the near future.[17]

Nuclear power is firepower at its zenith. India possesses nuclear weapons and has delivery means in terms of short, intermediate and intercontinental ballistic missiles. Further, there are aircraft capable of delivering nuclear weapons, and the INS *Arihant*, the nuclear powered submarine, has completed harbour trials and would be moving to sea trials. Possibly, the *Arihant* would be inducted by 2015, thereby completing the triad of nuclear delivery systems. India is reported to be holding between 80 to 100 weapons though no official announcements have been made by the Indian government on the subject. This is correlated with the weapons grade plutonium produced by the Bhabha Atomic Research Centre, Mumbai. India has a declared a nuclear no first use policy and is in the process of developing a nuclear doctrine based on credible minimum deterrence. The Indian National Security Adviser (NSA), Shri Shivshankar Menon, signalled a significant shift from "no first use" to "no first use against non-nuclear weapon states" in a speech on the occasion of the Golden Jubilee celebrations of the National Defence College on October 21, 2010. India is not a signatory to the nuclear Non-Proliferation Treaty (NPT) or the Comprehensive Test Ban Treaty (CTBT). It is a member of the International Atomic Energy Agency (IAEA) and six of its 17 nuclear reactors are subject to IAEA safeguards. By 2014, the number of reactors to be placed under IAEA safeguards would go up to 14.

The Nuclear Command Authority (NCA) of India is the nodal agency for all command, control and operational decisions regarding India's nuclear weapon stockpile. The NCA was established on January 04, 2003. It has a Political Council and an Executive Council. The Cabinet Committee on Security (CCS) which comprises the Prime Minister, Home Minister, Defence Minister and Foreign Minister forms the Political Council, and the Executive Council is chaired by the NSA. A Strategic Forces Command (SFC) was established with the responsibility for the management and administration of the country's nuclear stockpile. It was established concurrently with the NCA and has a Lieutenant General or equivalent from the Navy or IAF as its

Commander-in-Chief. The SFC is tasked to initiate the process of delivering nuclear weapons and warheads after acquiring approval from the NCA. The target area also will be selected by a process of discussion and final approval obtained from the NCA. The SFC has under its command all strategic forces and the communications network is geared up to maintain operational readiness. The SFC has delivery systems in terms of missiles and it is reported that it is planning to acquire 40 fighter planes capable of delivering nuclear weapons. Presently, the SFC depends on the IAF for aerial delivery of its nuclear assets.

Missile systems are historically associated with India. Missiles, as a constituent of firepower, were used by Tipu Sultan during the second Anglo-Mysore War in 1792, which resulted in 3,820 soldiers of the East India Company being taken prisoner. The rockets were deployed by Tipu's Army by special rocket brigades called *Kushoons*. These rockets were extremely effective and were reengineered by the British and used for a limited period. The present missiles held by the Indian armed forces are as a result of the Integrated Guided Missile Development Programme (IGMDP) which was ushered in by the erstwhile President of India, Shri APJ Abdul Kalam as Director, Defence Research and Development Laboratory (DRDL). After successfully guiding India's Satellite Launch Vehicle (SLV)-3 programme in the Indian Space Research Organisation (ISRO), he successfully developed the Akash, Prithvi and Agni missiles. Later, as Scientific Adviser to the Defence Minister, he established a joint venture with Russia to create a state-of-the-art cruise missile, the BrahMos.

The Akash is a medium range, surface-to-air missile with an intercept range of 30 km. The missile flies at a speed of around Mach 2.5 and can reach an altitude of 18 km. The system allows multiple targets to be attacked at the scale of four per battery. The missile is supported by a multi-target, multi-function phased array fire control radar, the Rajendra, with a range of 80 km in search, and 60 km in terms of engagement. It is guided by the radar and thereby it is difficult to jam. The system is meant for the Army and the IAF. The system is currently being inducted.

Ballistic missiles have proved to be the most successful part of the IGMDP. The Prithvi and Agni missiles have been developed and inducted into the armed forces. The Prithvi I was a single-stage, liquid-fuelled missile developed in 1990. The warhead payload weighed 1,000 kg; the range of the missile was 150 km and it was launched from transport launchers. The Prithvi II had a warhead of 300 kg and the range got extended to 250 km. This system was inducted in the Army and IAF in 2004. Further developments have enhanced the range to 350 km and the weight of the warhead has been enhanced from 500 to 1,000 kg. The K-15 Sagarika is a submarine-launched version of the Prithvi. This is a two-stage missile; the first is an underwater booster that powers the missile to 5 km above the surface of the ocean, and the second is a solid-fuelled stage with a thrust motor that propels the missile over 700 km. The Dhanush is a ship-launched variant of the Prithvi missile.

The Agni missile system forms a part of the Intermediate Range Ballistic Missile (IRBM) development. The system was first tested in1989 with a strategic payload of 1,000 kg. The Agni-I is a solid stage propellant which ranges from 700 to 800 km. The Agni-II is a two-stage solid-propellant missile with ranges between 2,000 to 3,000 km. The Agni-III is an IRBM, with two stages of solid propellant and ranges between 3,500 to 5,000 km. The Agni-I and Agni-II have already been inducted into the Services. In June 2011, it was reported that the Agni-III has been manufactured and inducted into the armed forces.

Further, an advanced version of the Agni-II has been improved and rechristened as the Agni-IV.

India's dream of joining the ICBM club was fulfilled by the immaculate launching of the Agni-V from Wheeler's Island at 0807 hours on April 19, 2012. The missile followed the designated trajectory and the three propulsion stages behaved correctly. The ships located in the path of the missile and at close proximity to the target tracked the vehicle and witnessed the final impact. The test saw the induction of the Ring Laser Gyro-based Inertial Navigation System (RINS) and Micro Navigation System and these enabled the missile to reach the designated target within a few metres. Precise guidance was provided by the high speed onboard computer with fault tolerant software. The missile had a launch weight of 50 tons and carried a dummy warhead of 1.5 tons. There will be six trial flights and the first flight, which will be a canister launch, will be undertaken in 2013. The missile is likely to be inducted into the Strategic Forces by 2015.

India, in conjunction with Russia, has undertaken joint development, production and induction of the supersonic cruise missile BrahMos. While the Prithvi and Agni are nuclear-tipped, the BrahMos carries a conventional payload. The missile can be launched from multiple platforms on land, sea, subsurface and air against land and sea targets. It has a maximum range of 290 km , a maximum velocity of 2.8 Mach and cruises at an altitude of 15 km. The BrahMos supersedes the most popular cruise missiles in the world by three times in terms of velocity and nine times the kill energy range. The missile has been inducted in the Army, on ships in the Navy and the development process is on for firing the BrahMos from fighter aircraft and submarines. DRDO is also developing a subsonic cruise missile, the Nirbhay, which would have a speed of .7 Mach and a range of 1,000 km. Recently, DRDO tested a 150-km ballistic missile, the Prahar, which has an accuracy of less than 10 m. The issue which merits attention is the fact that all Indian ballistic missiles, being nuclear tipped, a conventional ballistic missile like the Prahar, on being fired in conventional exchanges, could be mistaken for a nuclear tipped missile, thereby inviting a possible nuclear response. Apart from these missiles, DRDO is developing ballistic missile defence. This is a two-tiered system consisting of two interceptor missiles, namely, the Prithvi Air Defence (PAD) missile for high altitude interception and the Advanced Air Defence (AAD) missile for low altitude interception. The PAD, named Pradyumna, has a maximum interception altitude of 80 km and is capable of engaging ballistic missiles that range between 300 to 2,000 km at a speed of Mach 5.0. The missile system would be developed for intercepting weapon systems which range more than 5,000 km and fly at altitudes up to 150 km,

## Matching of Firepower Systems

The firepower assets of China and Pakistan are credible and India would politically try its utmost to avoid a two-front war. Mathematically, China would be applying the firepower of 30 divisions plus against 10 divisions of the Indian Army. This is an estimated figure, considering China's mobilisation capabilities and its political claims in terms of disputed territory. Quantitatively, China has an edge and what needs to be examined is our qualitative comparison of equipment which would cause a significant difference, particularly in the mountains. It is quality which India must focus on to turn the scales against the Chinese. In the initial stages of warfare, firepower was related to numbers and hand-held weapons. The greater the number of personnel, you could put into the attack, the greater your chances of winning the battle. Sun Tzu, as early as 500 BC, expressed issues differently in his treatise *The Art of War*. The most important was that large numbers are but lambs led to the slaughter unless they are correctly employed. This is particularly true of the mountainous terrain in our Sino-Indian and Indo-Pak borders where tactical skills combined with preponderance of firepower can decimate objectives, as was observed during the Kargil conflict of mid-1999. Most of Sun Tzu's thoughts are towards increasing firepower. In his own words, "The victorious strategist seeks battle after the victory has been won, whereas he who is destined to defeat, first fights and afterwards looks for victory."[18]

To rationally examine the firepower assets, we will first analyse aspects pertaining to artillery. The Chinese main field gun is the 122mm, with ranges of the D-30 at 15,300 m, the M 1931gun, Type 60 FG and self-propelled ranging 22,000 m. Pakistan has the 105mm M 101 with a range of 11,000 m, and 122mm D-30 which is similar to the Chinese equipment. The Indian Artillery has the 105 mm light field gun which has a maximum range of 17,400 m and a few regiments of the 122 mm D-30. Most of these guns would, however, be obsolete by 2030. Currently, the Chinese have their entire inventory based on a higher calibre which would have a greater payoff at the target end. Regarding the medium artillery, the Chinese have a few variants of the 130 mm, all of them ranging about 27,400 m, variants of the 152 mm ranging between 30 to 33 km and 155mm (45 calibre) with range going up to 21,000 m. The latest Chinese developments are the 155 mm SH-1 howitzer (52 calibre), PLZ 05 (52 calibre), PLZ 04 (54 calibre) with maximum range greater than 50 km. Both the PLZ versions are self-propelled, however, there would be no difficulty in producing a towed version. While the PLZ 05 is for export, the PLZ 04 is meant exclusively for use by the PLA. Pakistan has the 130 mm which is akin to the Chinese one. Further, in the 155 mm category, there is a variety of equipment. In the self-propelled version, there is the 155 mm M 109 A6 Paladin (39 calibre, range 39 km) and the Chinese Norinco 155mm SH1(52 calibre, range 50 km) 6X6 wheeled chassis. With regard to the towed guns, Pakistan has inducted the Turkish MKEK Panter (52 calibre, range 40 km with base bleed), 155 mm M114 (1,4600 m) and 155 mm M198 (39 calibre, 30,000 m). The Indian Army has the 130 mm which is similar to that of China and Pakistan. Further, 180 of these guns have been upgraded to the 155 mm (45 calibre) Soultam giving an addition range of 11 km. Apart from this, the Indian Artillery holds 410 pieces of the 155mm FH 77 (Bofors, range 30 km, 39 calibre). The artillery will be shortly equipped with 145 X 155mm 777 Ultra Light Howitzers

(ULHs) and trials have been completed for the 155 mm (52 calibre) wheeled SP gun. Currently, China and Pakistan have already inducted the 155 mm (52 calibre) weapon systems in their inventory, whereas India is primarily holding 155 mm in the 39 calibre which ranges about 10 km less than the modern 52 calibre weapon system. In addition, the Indian Artillery lacks a self-propelled gun which would have serious effects in a conflict with Pakistan in the plains sector and deserts. The Indian Artillery must take expeditious measures to speed up the modernisation process, particularly with respect to medium guns. As regards the rocket system, all countries have the equivalent of the 300 mm Smerch rocket system. However, China has a major advantage in that practically all its guns, rockets and ammunition are manufactured indigenously. As regards Surveillance and Target Acquisition (SATA) equipment, practically all three countries are at par. As artillery is a major component of firepower, there is a need for the Indian Artillery to expedite the procurement of guns and its associated equipment. At the present juncture, qualitatively, the artillery needs to catch up with our adversaries. Apart from this, the Indian Artillery must procure UCAVs and loitering missiles at the earliest to enhance its capability to undertake surgical strikes.

The next component we would match is firepower in terms of the three Air Forces. The 1991 Gulf War came as a wake-up call to the PLAAF. The Chinese leaders realise that nuclear deterrence alone is not enough when pitched against a state-of-the-art Air Force that could blunt the capability to retaliate. The PLAAF has modernised and developed aircraft with the assistance of Russia and Israel. China currently has the SU-27 which is in the same class as the F-15, the J-10 based on the Israeli Lavi fighter which would be rated at par with the F-16, the JF-17 which has been jointly produced with Pakistan and the fifth generation stealth fighter aircraft J-20 which has been test flown and is expected to be inducted by 2017. China also has the AN-12 modified for AEW&C. Pakistan's most capable fighter aircraft is the F-16.

There is a total of 22 F-16 As, 20 F-16 Bs, 12 F-16Cs and a few F-16 Ds. Further, two squadrons of an advanced variant of the J-10 are to be inducted into the PAF by 2015. Pakistan's F-7, Mirage-3 and Mirage-5 would be replaced by the JF-17 which is jointly developed by the Chinese and Pakistanis. Further, Pakistan has seven Saab 2000 AEW&C aircraft which have the Erieye radar with a range of 450 km. All in all, except for the F-16 and its upgrades, the PAF by 2030, would be primarily dealing with Chinese fighters. Chinese fighters are based on Russian and Israeli technology whereas Pakistan would have American and Chinese technology. On the contrary, the IAF, by 2030, would have a potent and mixed lot of fighters. The fighters comprise the Sukhoi 30 Mk 1, of which 165 are held and 107 are being manufactured by HAL. This is a multi-role air superiority fighter and compares favourably with respect to any aircraft held by the Pakistanis and Chinese. In addition, there are 67 MiG-29 fighters which are being upgraded to the MiG-29 SMT standard and 49 Mirage 2000 which are being upgraded. Our indigenous fighter, the Tejas, is being inducted and as of now, 40 aircraft are indented on HAL. Apart from these, 126 state-of-the-art Rafale MMRCA fighters will replace our ageing Jaguars and MiG-27s. The most significant change would be the completion of the development of the Indian FGFA by 2020 and the commencement of it induction by about 2022, with the total order of 250 being completed by 2030. The older aircraft of the IAF would be retired in a phased manner and this would see the MiG-21, MiG-29 and Mirage-2000 upgrades continue in service well into the next decade, with retiring of these aircraft possibly commencing in 2025. When we compare the fighter aircraft, we find that they are qualitatively at par, the Chinese and Pakistanis jointly owning many types of aircraft with Russian, French and Israeli technologies, whereas the Indian fighters are based on Russian and French technologies. Quantitatively, China has much greater numbers and there is a need for increasing our numbers by at least ten squadrons to ensure that we are not seriously disadvantaged in a two-front war. In a war with one adversary, our firepower with air assets would see us perform creditably. However, considering that our current versions have to continue for another 20 years, there is a definite need to increase, quantitatively along with quality. It a full spectrum conflict, numbers do play a role as both sides possibly have similar aircraft with BVR missiles, and pilot skills as well as network-centric capabilities would be extremely important in terms of AFNET with the Integrated Air Command and Control System.

As regards other issues, the Chinese have an airborne corps with airlift available for one division whereas by 2030, we would have, at best, capability to airlift about 2,500 troops. While the usage of these troops in Tibet would involve tremendous problems, there are suitable objectives for the Chinese all along the disputed areas for undertaking such operations. There are similar opportunities for us, particularly with the expected induction of the M 777 ultra light howitzer. The IAF is looking at a two-front war with an open mind and, for the first time, is planning to conduct a major exercise involving all five operational commands together. The exercise, named Exercise "Livewire", would be held in early 2013. It would test the IAF in operating different theatres simultaneously. The exercise would test all the five commands in their ability to respond to a collusive two-front threat from China and Pakistan. Overall, China has the quantitative edge and we must resolve this issue with deliberation at the earliest. It is pertinent to add that China, Pakistan and India already have UAVs and are developing UCAVs.

Matching the firepower of the Navies, one has to take into consideration that the PLA Navy is aiming for blue water capability which would possibly take place by 2030. Out of the three fleets, we are concerned with the South Sea Fleet as also naval assets in the Indian Ocean with bases at Coco Iislands in Myanmar, Gwadar in Pakistan and provision of facilities at ports in Maldives,

Sri Lanka and, possibly, Bangladesh. By 2030, the South China Sea and the Indian Ocean region would have a ship allotment which would include one aircraft carrier, two nuclear powered ballistic missile submarines (SSBNs), six nuclear powered submarines, about 30 other submarines, 10 destroyers, 12 frigates, 10 landing ships and about 100 patrol vessels to include mine-sweepers and auxiliary ships. This combined with the naval aviation would possibly be equipped with two to three squadrons of Sukhoi-30 and Sukhoi-33 as also about 20 squadrons of fighter bombers capable of attacking land and sea targets with two squadrons of maritime reconnaissance aircraft and an abundance of helicopters. These naval assets would help China to dominate the Indian Ocean and suitably fire cruise missiles from its ships on both land and sea targets, ably supported or assisted by the Pakistan Navy which, even in 2030, would be a small Navy based on a few frigates, destroyers and submarines. The Indian Navy would be a force to contend with by 2030. Based on the procurement process, the Navy would have possibly two, may be, three aircraft carriers, six nuclear powered submarines, a large number of other submarines and numerous frigates, destroyers as also other ships with C⁴I²SR systems which can match the Chinese naval might at standoff ranges. Of the three Services, the Indian Navy, with its well planned acquisition programmes, would be able to match the Sino-Pak firepower in 2030. Currently, India has the edge.

Missiles form an important component in the firepower inventory. In the current Libya conflict, the bugle was sounded by more than 100 Tomahawk cruise missiles landing in erstwhile President Gaddafi's compound. The entire conflict was won by missiles, rockets and air-to-ground laser guided bombs. China's Second Artillery, which handles conventional and nuclear missiles, is a force which excels in professionalism and there are 38 missile units. As stated, China has modernised its entire missile inventory. The main missiles which China would possibly use against India would be the ballistic missile DF-21A with a range of 2,150 km and a series of cruise missiles which are the SY-1, HY, FL, YJ, C 701, C801; all these range between 300 to 600 km. Chinese missiles can engage every part of India and considering its policy of no first use of nuclear weapons, it would be in a full spectrum conflict initially using conventional missiles. The range of China's conventional missiles is phenomenal and covers the entire Indian subcontinent. Pakistan's missile programme is Indo-centric, with assistance from China and North Korea. Pakistan would be keen on early usage of nuclear weapons on military targets to deter India from adopting a proactive mode of undertaking operations. The Indian missile programme has narrowed the strategic gap with the successful firing of the Agni-V but we have only the BrahMos which can fire conventional warheads up to a range of 290 km. DRDO is developing a subsonic cruise missile, the Nirbhay, which would have a range of 1,000 km. This is in the development stage and is likely to have its first firing by the end of 2012, possibly early 2013. A minimum of six firings will have to be done before induction; therefore, it would be around the first half of the next decade that the missile would be inducted, provided all goes well and DRDO is able to meet the timelines. A conflict with China may be purely with missiles and, possibly, air. In such an eventuality, we have no chance of engaging Chinese missile sites or their storage locations at Delingha and Kunming as they are out of the range of our conventional systems. On the ICBM front, the Chinese have the MIRVs, which give them tremendous advantage, especially against BMD.

Firepower in 2030 will include activities in outer space. Currently, the $C^4I^2SR$ of China, Pakistan and India is heavily dependent on outer space. China launched its navigation satellites, known as the BeiDou system comprising 35 satellites, on April 13, 2007, and the entire system, would be operational by 2020. By 2020, China be capable of launching its weapon systems at any point on the globe, using its navigation system, with accuracies possibly of one to two metres. China also successfully fired an ASAT weapon on January 17, 2007, and is currently examining interference with existing satellites to degrade military communications. The Indian armed forces are currently dependent on national assets to cater for their $C^4I^2SR$ needs. India needs satellites dedicated to its Space Force Command and its own navigation system satellites. Overall, China definitely has a qualitative and quantitative edge over India. Pakistan is totally dependent on China for its firepower procurements and technological advancement, including nuclear weapons.

## High Technology Weapons of PLA to be Inducted Possibly by 2030

It is an age old practice in Chinese history to seek a trump card or assassin's mace weapons against an adversary. Information about the development of assassin's mace weapons is classified and, in the present context, forms a part of the asymmetric weapon development. Impetus for development of such weapons was given by erstwhile President Jiang Zemin. It is important to note the statement of Maj Gen Wang Hongguang, Commandant of the PLA Armoured Forces Engineering Academy, who stated in January 2001, "Extending our vision to the 21$^{st}$ century, the extensive application of information technology, nano technology, new materials technology, new energy resources technology and other high and new technologies will enable the PLA to be reborn. Operational space will become even wider, operational modes will become more varied, response time will become even quicker, actions will be more agile and attacks will become more forceful. All types of weapon systems, support and logistics systems will combine with the information flow to become one entity, implementing real coordination of high efficiency and accuracy, real-time attack and real-time support..... Electromagnetic artillery, kinetic bombs of high altitude and high speed, and smart weapons and high efficiency pulse weapons with laser and particle beam capabilities will, through their unique capabilities, release the operational capabilities and threat capacity of the ground forces. The mass tactics of larger infantry operations will only remain in the people's armies."

In terms of firepower, lasers, radio frequency weapons, thermobaric weapons, hypersonic vehicles, electromagnetic weapons, nano weapons, biotechnical weapons, stealth as also counter-stealth weapons and supercavitating underwater weapons are being developed by China.

China is developing lasers for weapons, communication and radar applications. As per the US Department of Defence, China possibly has the capability to damage under specific conditions the optical sensors on satellites that are vulnerable to lasers. Further, it is reported that the focus is to utilise lasers for destruction of personnel, precision-guided munitions, air defence and satellites. In addition, lasers are being developed to arm UAVs, satellites as also cannons and are being tried out for underwater communications. The Chinese have been contemplating basing lasers in outer space for degradation and destruction of satellites.

Taiwanese sources have confirmed to Mr Richard D Fisher Jr who has authored the book *China's Military Modernisation, Building for Regional and Global Reach* (2008) that the PLA was nearing the deployment of a new non-nuclear Electro-Magnetic Pulse (EMP) warhead on a Short Range Ballistic Missile (SRBM). This is a High Power Microwave (HPM) warhead fitted on to a new version of the SRBM. EMP and HPM weapons are radio frequency weapons used for paralysing micro circuits, computers, radars and other sensors in all communication systems. These could be effective against ships and land systems. The PLA has also developed thermobaric weapons for its artillery and Air Force. The development has taken place with help from Russia. The PLA is on the way to develop hypersonic vehicles, like the United States and Russia. These vehicles can also serve as low earth orbit launch vehicles and could be considered as the successor to the ICBMs. The Chinese are developing these vehicles with Russian assistance.

China is possibly using Denel technology to build the Dark Sword UAV with the Shenyang Aircraft Corporation. China has shown great interest in electro-magnetic weapons. These weapons use magnets and chemical propellants to give projectiles speed and range far greater than is possible with present-day propellants. Such equipment will enable longer ranges and high speed to projectiles to engage targets as also counter other projectiles.

The PLA is focussing on nano technology with a view to transform military technology. Presently, there are more than 100 projects employing around 3,000 engineers. The projects include micro weapons, nano satellites and military equipment. China is also dedicating numerous projects to stealth, counter-stealth and bio-technology weapons. China has also procured supercavitating underwater weapons from Kazakastan. Supercavitating underwater weapons move at high speeds, on occasions faster than the speed of sound. They could be used as torpedoes, anti-torpedoes, anti-mines or missiles. Sources indicate that China is developing these weapons and could induct the same once development is complete.[19]

China, after the Gulf Wars, has modernised the PLA in its firepower assets and adopted changes in the art of orchestrating a conflict. Its infusion of high technology in the fighting elements has revolutionised the entire concept of conflict by a system of system approach which enables victory in war by creating asymmetries of warfare.

## Notes

1.  William Magioncalda, "A Modern Insurgency: India's Evolving Naxalite Problem," *South Asia Monitor*, No 140, April 08, 2010 (Washington DC: Centre for Strategic and International Studies).
2.  Ashley J Tellis, "Policy Outlook, March 2012".
3.  Col G D Bakshi, VSM, "The Sino Vietnam War 1979. Case Studies in Limited Wars," *Bharat Rakshak Monitor*, Vol 3(3), November-December 2000, p. 3.

4. Harinder Singh, *Pakistan 2030: Possible Scenarios and Options, Asia 2030* (New Delhi: IDSA), pp. 122-123.
5. "China May Resort to Grab Indian Territory, Expert Report," www.dnaindia.com/world/report,1656593.
6. Col R S N Singh, *China, Asian Strategic and Military Perspective* (Observer Research Foundation, Lancers Publishers and Distributors 2005), pp. 42-53.
7. Air Cmde Ramesh. V Phadke, "Defending Indian Skies Against the PLAAF," *Indian Defence Review*, January/March 2012, pp. 39-45.
8. Robert S Norris and Hans M Kristensen. "Chinese Nuclear Forces 2006," http:// the bulletin.metapress.com/content/1wo35m8u644p864u.
9. William Atkins, "Chinese BeiDou Navigation Satellite Launched from Long March 3 A" www.itwire.com/science-news/space/9201-chinese-BeiDou-navigation-satellite-launched-from-long-march-3a-rocket.
10. "Pakistan Land Forces," *Military Technology*, January 2011, p. 366.
11. "Pakistan Navy," http://www.paknavy.gov.pak. Pakistan Navy website on submarines http://www.paknavy.gov.pk/xcrafts-html, April 09, 2012.
12. *Pakistan, Brahmand World Defence Update 2012* (New Delhi: Pentagon Press), pp. 361-362.
13. "Pakistan Military Consortium, RBS 70, ANZA Mk I and Mistral," http:// www.pakdef.info/pakmilitary/airforce/sam/bofors.html.
14. *India, Brahmand World Defence Update 2012* (New Delhi: Pentagon Press), pp. 213-214.
15. n. 10, p. 342.
16. Ibid., p. 344.
17. Ibid.

# OPTIMISATION OF FIREPOWER ASSETS IN A CONFLICT WITH CHINA OR PAKISTAN

*The longer the fighting, the harder the war, the infantry and, in fact, all arms lean on the Gunner.*

— Field Marshal Bernard Montgomery,
Viscount of Alamein

## Introduction

China, for the visualised future, will remain an important strategic challenge for India. It is a country whose geo-politics directly affects India.[1] China is an emerging superpower and India is a powerful country in South Asia. India and China have problems on the border issue. While China has recognised the McMahon Line with Myanmar, the same does not apply for India. The McMahon Line is named after Sir Henry McMahon, Foreign Secretary of the British run Government of India when the treaty was signed between Britain and Tibet at Simla in 1914. China rejects the Simla Accord, contending that the Tibetan government was not sovereign and, therefore, treaties concluded by it were not valid. As on date, Chinese maps show Arunachal Pradesh, Aksai Chin and Shaksgam Valley, known as South Tibet, as a part of China. China has been in illegal occupation of 43,180 sq km in Ladakh sector and 5,180 square sq km of Shaksgam Valley. China alleges that India is in illegal occupation of 90,000 sq km of Chinese territory.

During the Cold War, the strategic scenario of South Asia was pentagonal.[2] India and Pakistan were the two directly affected countries. The US, erstwhile Soviet Union and China were the other countries involved in the region. All of them influenced activities in this area but with the end of the Cold War, the role of Russia has considerably reduced its role, leaving the subcontinent in the hands of India, Pakistan, the US and China. The major issue between India and China is the border dispute; surprisingly, China has resolved its border with numerous countries. In the 1960s, China resolved borders with five countries. These were Myanmar (1960), Nepal (1963), North Korea (1964), Mongolia (1964) and Pakistan (1965). As time has progressed, China has resolved boundary issues with six more countries: Laos (1993), Vietnam (1999), Russia (2004), Kazakhastan (2002), Tajikistan (2002) and Krygyzstan (2004). Out of the 14 countries that share borders with China, 12 countries have resolved their disputes – the only two left are India and Bhutan. It is pertinent to note that Bhutan's peace treaty with India of 1949 stipulates that Bhutan would be guided by the advice of the Government of India with regard to its external relations. Though China has recognised Sikkim as a part of India in exchange for India's acknowledgement of China's sovereignty over Tibet in July 2006, currently both sides have stuck to their guns and settlement of the dispute appears to be taking a long time.[3] It may be pertinent to note that Chinese Prime Minister Wen Jiabao, during his last visit to India, stated that it would take a long time to settle the boundary issue. This is a departure from the earlier position and India has to improve the infrastructure along the border expeditiously.

The other critical issue is China's supply of nuclear technology to Pakistan for its nuclear weapons programme. As on date, China, Pakistan and India are nuclear weapon states. Pakistan's programme is India-centric while India's programme caters for both China and Pakistan. Both these issues are volatile and could easily lead to confrontation. [4]

With Pakistan, the problems are the land disputes of Kashmir and Siachen as also the maritime dispute of Sir Creek. The major issue is the terrorist infrastructure which exists in Pakistan and its utilisation for attacking targets all over India. Any irresponsible terrorist attack may compel a military response from India which could exacerbate into a sub-conventional conflict or into a full spectrum conflict.

## Likely Nature of Conflict

It is extremely difficult to correctly assess the Chinese military intentions. China's stated operational strategy guideline is called active defence. This stipulates not to initiate conflicts or undertake aggression.[5] In 2007, Lt Gen Zhang Qinsheng, PLA intelligence chief, stated that strategically, China adheres to defence, self-defence, and would win by striking only after the enemy has struck. He was emphatic that China would not fire the first shot. It is interesting to note that PLA officers are taught the reverse of what the intelligence chief stated. The PLA National Defence University text-book, *The Science of Campaigns*, says, "The essence of active defence is to take the initiative and to annihilate the enemy." [6] Details obtained from another text-book *The Science of Military Strategy* elucidate, under high-tech conditions for the defensive side, the strategy of gaining mastery by striking only after the enemy has struck does not mean waiting possibly for the enemy's strike. Thereafter, it modifies the definition of first shot by stating that if hostile forces such as religious extremists, national separatists and international terrorists challenge a country's sovereignty, it could be considered as firing the first shot on the plane of politics and strategy.[7] Accordingly, a war could commence against Xinjiang, Tibet and Taiwan by this logic. Thus, a preemptive strike against Taiwan, Xinjiang and Southern Tibet (Arunachal) is permitted. Therefore, we must be prepared to face a full spectrum conflict against a nuclear backdrop from China.

Currently, the PLA has incorporated integrated joint operations amongst the three Services based on the lessons learnt during the Gulf Wars, Afghanistan, Bosnia and the current conflict in Libya. The degree of PLA success in these operations is difficult to judge from sources in the open domain. However, the importance given to the subject can be observed by the fact that after 2004, Commanders of the Navy, Air Force and the Second Artillery have become permanent members of the Central Military Commission high command.

With these changes, the PLA's current task in war is to attack and destroy the enemy's most powerful critical assets. Active defence has graduated with the present doctrinal progression to fighting local wars under informationised conditions. The current procedure entails offensive actions unlimited by time and space once hostilities have begun. The focus is on mobility and state-of-the-art electronics and weapon systems. With the advent of the new strategy of operations, PLA ground forces are training in combined arms and integrated joint operations. Special emphasis is being laid on airborne assault, close air support, reconnaissance, electronic warfare and cyber operations. The ground forces appear to place heavy emphasis on using conventional missile strikes and special forces to deliver possibly preemptive strikes against the enemy's command and control centres as also logistics areas. These would be followed by deployment of combined arms forces which would rely on area denial by the PLAAF. It is extremely interesting to note that the PLA has begun moving away from scripted exercises towards greater free play, with an emphasis on lessons learnt, rather than on smooth victory by the PLA.

China has evinced great interest in the Indian Ocean and by 2030, the Chinese PLAN would be actively involved in this area. The PLAN is currently focussed on acquiring dominance in the Yellow Sea, Taiwan Strait, East China Sea and South China Sea.

The Indian Ocean comes second in the present order of priority. Though China would like to move to the Indian Ocean, there are numerous issues engaging China in the first tier, The US decision to retain strong forces in the Asia Pacific, a strong Japanese naval power projection, and development of the naval capabilities of Vietnam, Indonesia and Australia would possibly delay Chinese deployment in the Indian Ocean region. However, by 2030, the PLAN would be active in the waters of the Indian Ocean. The Indian Navy will be challenged in its own backyard in another 15 years.

The PLAAF, like the other two Services, has become more offensive and is developing the capability to use its firepower assets around China's periphery with modern aircraft. The Chinese Air Force is focussing on $C^4I^2SR$, Airborne Early Warning (AEW) and mid-air refuelling. With regard to nuclear weapons, China has concentrated on improving the credibility of its nuclear deterrence. This has resulted in the Second Artillery moving from liquid-fuelled rockets to solid fuel and from silo-based missile towards mobile launchers and reloadable launchers. Since the last decade, the Second Artillery has been developing conventional missiles to enhance the PLA's fighting capability in a conventional war under a nuclear shadow. China is keen on using space for military operations. Though no official space doctrine is available in the open domain, both the Chinese Air Force and Second Artillery are involved in military applications of outer space.

Pakistan's strong friendship with China helps it against the adverse fallout from its anti-India policies. Further, the United States still looks at Pakistan as a major non-NATO ally and is keen to scrounge on whatever assistance can be obtained from Pakistan rather than seriously try to resolve the cross-border terrorism issues between Pakistan and India. All the organs of Pakistan, be it the political leaders, the Army, the Inter-Services Intelligence (ISI) and the bureaucracy believe that it is only cross-border terrorism that compels India to engage with Pakistan. It is important to note that Pakistan's possession of nuclear weapons imposes constraints on India's countervailing strategy. Pakistan strongly feels that its nuclear capability prevents India from initiating a war which possibly would escalate to the nuclear level.[8]

Viewing all these issues, as also the existing problems of India with China and Pakistan, a conflict is possible if the situation goes out of control. Pakistan is fully involved in trans-border terrorism which meets its strategic requirements. While China's main focus is on Taiwan, it is deeply concerned about Tibet. China could involve itself against India in a Low Intensity Conflict (LIC) or a full spectrum conventional conflict under a nuclear overhang. An LIC may involve only cyber war in which China knocks out the important command and control nets of our armed forces. As our firepower on land, the sea and in the air is dependent on communication networks, the entire weaponry employment could be degraded by cyber warfare. It is reported that China is covertly aiding militant groups in the northeastern portion of India who are able to undertake sporadic missions against the Indian security forces. Of late, the good relations of India with Bangladesh and Myanmar have reduced the impact of these insurgent elements considerably.

China would undertake a full spectrum conventional conflict under a nuclear overhang to either capture Tawang and other important areas of Arunachal Pradesh and Ladakh or to teach a lesson to India as it did with Vietnam during the 1979 War. Firepower will play a major role in this conflict.

### Firepower in a Full Spectrum Conflict

War with China will be fought under a nuclear overhang. India and China have a nuclear doctrine of 'no first use' which could make decision-makers on both sides cautious. China could launch a major offensive in Arunachal and Ladakh with the purpose of capturing Tawang and grabbing areas in Arunachal and eastern Ladakh. While there are other options also, this is possibly the most likely one. The force level estimated, based on its objectives, would be about 30 divisions plus with elements of the Second Artillery. In the most likely option, China would broadly launch the main offensive on the Kameng division with about eight to ten divisions, on the rest of Arunachal Pradesh with about six to eight divisions, on Ladakh with six to eight divisions, and holding attacks against Sikkim with four to five divisions as also on the UP-Tibet border with two to three divisions, and 36 Sector with one division. There would also be a reserve of three divisions. Operations would be integrated joint operations with possible use of airborne elements. The forces would mainly comprise about nine divisions from the Chengdu Military Region, six divisions from the Lanzhou Military Region, about 12 divisions from other Military Regions, six divisions from the Tibet Autonomous Region and elements of 15 Airborne Corps.

The conflict would be preceded by disruption of communication networks by cyber warfare. There is a possibility of China destroying our remote sensing satellites to avoid correct recognition and detection. China would use the DF-21 ballistic missile which has a range of 2,150 km as also cruise missiles which range from 300 to 600 km and would be used on command and control centres, logistics installations, bridges, mountain passes, airfields, UAV sites, missile storage locations and gun positions. China would also have ABM systems to counter our missiles. The air power available to China for these operations will operate from the five airfields in Tibet as also Chengdu and Kunming. Air availability would comprise about 40 fighter squadrons which would include by 2030 at least eight squadrons of the J-20 stealth fighter aircraft. Aircraft would also be used for close air support of ground troops as also airborne operations. The PLA has recently practised airborne operations in training exercises in Tibet and would definitely use this as a *coup de main* operation. The recently conducted exercises by the PLA in Tibet Autonomous Region (TAR) by the Lanzhou and Chengdu Military Regions in July-August 2012, included missile tests, use of fighter jets, the airborne corps and ground troops for the capture of passes at 5,000 m Above Mean Sea Level(AMSL). It is pertinent to note that passes at these altitudes are in the Greater Himalayas on the Indo- China border, so the PLA would have tremendous firepower for the offensive. There would be about 30 (plus) divisional artillery brigades, each having four field regiments and one medium regiment,which works out to a total of 120 field regiments and 30 medium regiments. There would be about six Group Army Independent Artillery Brigades each having four medium regiments and one rocket regiment which works out to a total of 24 medium regiments and six rocket regiments. In addition, there would be two artillery divisions each with nine medium regiments and three rocket regiments. Accordingly, there would be a total of 120 field regiments, 72 medium regiments and 12 rocket regiments for the offensive. Our own firepower resources currently have numerous voids and would require a proactive attitude by the Ministry of Defence, and Army and Air Force authorities to make up the operational voids so as to be prepared with our present capability in the mountains. By 2030, we would need to carry out a relook at our entire force levels which currently are inadequate to meet an onslaught of 30 divisions plus on the Sino-Indian border. Viewing our present force structure, we are heavily deployed against Pakistan with our three strike corps poised for a shallow offensive against a nuclear overhang. For a dissuasive capability vis-a-vis China, we would need possibly two strike corps for the mountains, with two artillery divisions and FGFA, Rafale jet fighters and Sukhoi 30 jets to be deployed at our forward airfields which would be able to engage the Chinese aircraft and assaulting forces effectively while they are in the process of mounting the offensive.

There are hardly 18 years to go for 2030 and the artillery of the Indian Army needs to innovate, as despite India's best efforts, procurements take a long time to fructify as all decisions are collegiate which require vetting by numerous bodies. We are aware that

Pakistan's focus is on cross-border terrorism and any conventional offensive by Indian forces would possibly lead to usage of nuclear weapons. Accordingly, the artillery of the strike corps, particularly those with the artillery divisions and a few medium regiments need to be dual tasked for the mountain regions bordering China. China itself has only 50 percent of the forces coming from the Chengdu and Lanzhou Military Regions whereas the remaining forces are drawn from other Military Regions not directly positioned against the Sino-Indian border. Further periodic exercises must be held of these formations in the designated area of operations on the Sino-Indian border. Apart from the yearly operational alerts, two-sided exercises must be conducted in conjunction with the Indian Air Force once every three years in which all the commands must participate. The exercise being an inter-Service one, must involve the Ministry of Defence and should have its focus on China with a collusive effort from Pakistan.

With regard to weaponry, our reach in the mountains is about 40 km which is the range of the Bofors gun in high altitude terrain. This just serves the purpose of the contact battle, while the requirement is of taking on the adversary right from his mobilisation and concentration areas. For the intermediate battle, we must use rockets whose shock action will break the adversaries' will to fight. Currently, none of our rocket regiments is deployed in the mountains despite their employment during the Kargil conflict in 1999. The upgraded Grad BM-21 with the extended range, the Pinaka multi-launcher rocket system and the Smerch, with fewer tubes mounted on the Kamaz vehicle, would be effective against the Chinese in high altitude terrain, with maximum ranges varying from 40 km, 37 km and 90 km respectively, Efforts must be made to procure the Smerch with fewer tubes mounted on a Kamaz vehicle for the mountains. Further, it is reported that the Russians have developed rockets with ranges of 120 km for the Smerch system. We must not only procure the rockets but obtain the Transfer of Technology

(ToT) for the entire variety of Smerch ammunition. In view of the asymmetry of firepower, there is a necessity for acquiring five additional Smerch regiments for deployment of a regiment each in eastern Ladakh, the UP-Tibet border, Sikkim, Kameng division and the rest of Arunachal Pradesh. It is obvious that the Heron UAV as also the future DRDO UAV Rustam, with capabilities of operating at 30,000 ft, would be able to undertake detection, identification and recognition as also Post Strike Damage Assessment (PSDA).

A full spectrum conflict with China would be in the conventional domain as both countries have pledged no first use of nuclear weapons. Missiles capable of firing conventional warheads would have their might displayed with full fury. The Chinese have the DF-21 mounted on a Transporter Erector Launcher (TEL) with a range of 2,150 km and numerous cruise missiles with ranges from 300 to 600 km covering the entire Indian subcontinent. The missiles could be deployed south of Lhasa in Tibet or in Delingha (Qinghai) and around Kunming in the Sichuan province. Currently, the only missile with conventional capabilities held by the Indian Army is the BrahMos supersonic cruise missile with a range of 290 km. Our targets in Tibet are more than 500 km, and to engage missile storage locations in Qinghai or Sichuan province as also important communication centres and command and control centres, and the airfields of Kunming and Chengdu, we would need missiles with a range of 1,000 km. DRDO is developing the subsonic cruise missile, Nirbhay, which will possibly have its first development flight by the end of 2012 or some time in 2013. If all goes well, it would take possibly another 10 years before the missile is inducted into our armed forces – possibly by 2022-25. However, there are missiles which are yet to be inducted after 29 years of development. Accordingly, there is a need for contingency planning. We could plan on one of the

Agni missiles carrying conventional warheads or extending the range of our present supersonic cruise missile.

We also need ABM systems which would be able to tackle the Chinese missiles. The Chinese are developing ABM systems similar to the Russian S-300 and S-500. Currently, we are going in for the Akash with a range of 25 km and DRDO has undertaken development trials for the PAD and AAD systems. Our tests have been carried out primarily with a modified Prithvi missile from Chandipur to simulate an enemy missile and an interceptor AAD missile from Wheeler's Island in the Bay of Bengal destroyed the incoming missile. Similar tests were undertaken by the PAD missile using the Swordfish Long Range Tracking Radar (LRTR). This two-tiered system, when inducted, will be able to intercept targets up to 5,000 km range but while we are finalising these systems, we should try and get a state-of-the-art system from the USA as we have agreed to cooperate in the field of missile defence. Possibly, this will require bilateral agreements similar to the US-Israel and US-Japan agreements on ABM. We could consider the NATO Active Layer Tactical Ballistic Missile Defence (ALTBMD). This system integrates dissimilar legacy systems and upgrades air command and control systems, resulting in the selection of the correct interceptor for the task of destroying the hostile missile.[9]

Prior to launching an offensive, China would like to possibly deactivate our surveillance systems by knocking down a few of our remote sensing and communication satellites using ASAT missiles which have been successfully tested. Our own developments would take some time and there is a requirement to push our resources to develop this capability expeditiously.

China would employ its Air Force for firepower dominance in the area of operations. Recent exercises held in Tibet by the Chengdu Military Region witnessed the use of fighter aircraft from Tibetan airfields as also elements of the airborne corps. Currently, our Air Force has commenced deployment of our fighters in airfields close to Arunachal and other areas to effectively counter this threat. Our aircraft should be able to attack Chinese airfields, missile storage locations, detraining stations in Tibet as also the airfields at Kunming, Chengdu and missile storage locations in Qinghai province. Our Air Force must be able to provide strategic lift to our troops and firepower assets to enable deployment or changes in application of these weapons based on the development of operations. Our Air Force is correctly focussing on strategic lift capability in terms of both aircraft and helicopters. These aircraft and helicopters are being procured and it can be safely predicted that these would be inducted by 2030. These airborne firepower platforms need to be fine-tuned for operations in the high altitude regions of Tibet. Our visualised modernisation of airlift and lighter weapons airborne operations, with dedicated artillery support, would be required for operations in Tibet. The Army and Air Force must synergise their operations to ensure that these are possible in high altitudes. Nano technology and its pragmatic application would result in lighter and smaller equipment which would exponentially assist our operations in high altitudes. DRDO must make forays with collaborative research in this field of technology.

Air operations in the ensuing decades would depend on large scale employment of UAVs, UCAVs and other Unmanned Aerial Systems (UAS). While UAS covers the entire gamut of unmanned systems, UAVs would play a predominant role with regard to surveillance and UCAVs would be a combination of surveillance, target acquisition, engagement, PSDA and destruction. Providing firepower for our aerospace land battles in Tibet it would be essential to gain mastery of operating these systems in high altitudes. The Chinese are developing a large number of UAS and would be utilising the same in operations in Tibet. Our Aeronautical Development Establishment is in the process of developing a UCAV known as the Aura which is in the preliminary stages. As the process would take considerable time, it is essential that the Indian armed forces produce a UCAV which could be gainfully employed in the air-land battle over Tibet.

As in the case of Tibet, the naval threat is assessed by examining capabilities and intentions. The Chinese maritime strategy calls for its naval assets to undertake missions pertaining to protection of vital Sea Lanes of Communication (SLOCs) running across the Indian and Pacific Oceans, enforcement of Chinese territorial claims in the East and South China Seas, deterring and preempting Taiwanese moves towards independence and maintaining a credible underwater nuclear deterrent against the US, Russia and India. On the maritime front, currently China has opposed the oil exploration by ONGC Videsh Ltd in the South China Sea and is busy establishing itself in the Indian Ocean. India has decided to ignore China's opposition and interference in its moves to undertake oil exploration, stating that these are international waters. In the Indian Ocean, China has a base in the Coco Islands, and has established ports in Sri Lanka, Maldives and Gwadar in West Pakistan. China is extremely keen to defend its SLOCs which pass from the Gulf through the Indian Ocean to its ports in eastern China. China has offshore gas assets near Sittwe in Myanmar. A gas pipeline has possibly been completed between Sittwe and Kunming. The Indian Navy with its modernisation plans is well poised to undertake the Chinese threat in the Indian Ocean but certainly lacks the ability to operate beyond the Malacca Strait based on current acquisitions and berthing facilities. For operations in the South China Sea, the naval and firepower assets of the Vietnamese need to be utilised with the collaborative efforts of other countries. It is reported that the USA has been offered berthing at Cam Ranh Bay in central Vietnam from where it would be possible to contest Chinese naval threats in the South China Sea. As regards maritime firepower capabilities, currently India matches China in the Indian Ocean but it is a matter of time before China's indigenous aircraft carrier is operationalised and then possibly the Indian Navy would have to ensure that its modernisation process is on track to undertake its task from the Gulf of Aden to the Strait of Malacca.

Firepower in a full spectrum conflict would play a predominant role and by the current pace of development, China has a definite edge with regard to the Army, a balance with regard to the Air Force and Navy, and preponderance with regard to missiles as well as nuclear power. As regards fighting a two-front war, considering its firepower assets, it is in India's interest that it should possibly be avoided. In an exclusive war with Pakistan, the firepower assets of India outnumber those of Pakistan and this would compel Pakistan to adopt a policy of use of nuclear weapons at an early stage to prevent India from adopting a proactive stance. Victory in a conflict in 2030 will be decided by the asymmetries of firepower. Accordingly, it would be necessary to modernise our forces expeditiously.

## Notes

1. Sunil Khilnani, Pratap Bhanu Mehta, Lt Gen Prakash Menon (Retd), Nandan Nilekani, Srinath Raghavan, Shyam Saran, Siddharth Vardarajan, "China, Non-Alignment 2.0, 2012," p.7.

2. Stephen. P Cohen, "Geo-Strategic Factors in India Pakistan Relations," *Asian Affairs*, Vol 10, No 3, Fall 1983, p.28.

3. George G Gilboy and Eric Heginbotham, *Foreign Policy: Use of Force and Border Settlements, Chinese and Indian Strategic Behaviour, 2012* (New Delhi: Cambridge University Press Private Ltd), pp. 84-85.

4. J K Boral, "Conflict and Cooperation in India-China Relations," *Journal of the Defence Studies*, Vol 6. No 2, April 2012 (New Delhi: Institute for Defence Studies and Analyses), pp. 82-84.

5. Office of the US Secretary of Defence, "Annual Report to Congress, Military Power of the People's Republic of China 2007," p.12

6. US Department of Defence, "PLA Report 2007," pp. 12-13.

7. Peng Guangqian and Yao Youzhi, *The Science of Military Strategy*, English First Edition (Beijing: Military Science Publishing House, 2005), p.426.

8. Sunil Khilnani, Rajiv Kumar, Pratap Bhanu Mehta, Lt Gen Prakash Menon (Retd), Nandan Nilekani, Srinath Raghavan, "Pakistan, Non-Alignment 2.0," 2012, p.12.

9. Alan H Merbaum, "Integrated Air and Missile Defence Architecture for India," *Strategic Affairs*, May 2008, pp. 22-24.

# MODERNISATION

$6$

*If you know your enemy and know yourself, you would not be imperilled in a hundred battles; if you do not know your enemy but know yourself, you will win one and lose one; if you do not know your enemy nor yourself, you will be imperilled in every single battle.*

— Sun Tzu

## Need for Expeditious Modernisation of Firepower

China by its build-up of Comprehensive National Power (CNP) would like nations to submit to its claims rather than get involved in fighting. Sun Tzu, the great philosopher, has stated that to subdue the enemy without fighting is the acme of skill. China, apart from modernising the PLA exponentially, has undertaken development of infrastructure in Tibet to mobilise its firepower assets expeditiously. The rapid development has enabled movement of missiles, firepower weaponry and ammunition from other Military Regions in possibly a month, which previously would have taken two seasons. Presently, there is an extensive road network of over 40,000 km and a 1,145-km railway line from Gormo in Qinghai province to the Tibetan capital, Lhasa.

Further, a pipeline has also been constructed from Gormo to Lhasa. The three highways, the Western Highway, connecting Lhasa to Kashgar, the Central Highway connecting Xining to Lhasa and the Eastern Highway connecting the Military Region at Chengdu to Linzhi are linked by smaller roads connecting every post by road along the Sino-Indian border. Further, the railway network is being extended and the Chengdu-Lhasa link would be ready in about five years. China has also built five airfields in Tibet and elements of the airborne corps have recently undertaken an exercise in Tibet. The Chinese have an excellent communications set-up with satellite communications backed by optical fibre.

China has state-of-the-art weapons which include cyber weaponry, ASAT missiles to knock out our satellites, $C^4I^2SR$ systems, ballistic and cruise missiles, both conventional and nuclear, to cover our entire landmass and island territories, space-based systems, preponderance of firepower from the air and ground as also about 30 divisions to be launched for operations from eastern Ladakh to Arunachal Pradesh. In such a scenario, our forces, though fewer in numbers, have the capability to dissuade China, if modernised expeditiously with state-of-the-art weapons. Further, our infrastructure along the Sino-Indian borders needs to be developed.

India specially needs to note the recent developments in the Chinese strategic missile launch from mobile launchers. The Second Artillery Corps comprises the ground-to-ground strategic nuclear forces, conventional operational tactical missile forces and support units. China today possesses a big family of various short range, medium range, long range and intercontinental range missiles. Further, China's missiles have become more powerful, accurate and faster, and practically all the missiles can be launched from vehicle mounted launchers and can accurately hit targets.[1] There is a definite need for India to take steps to procure equipment needed for modernisation at a rapid pace.

## Expeditious Procurement

On September 04, 2012, the Chinese Defence Minister, Gen Liang Guanglie, with a delegation of 23 members visited India and had discussions with the Indian Defence Minister. The two Defence Ministers were meeting after eight years and Gen Liang also gave a written reply to the editors of *The Hindu*, an important daily newspaper, on the border issue. He said, "The boundary issue in the China-India relations is an issue left over from history. It is also an issue at the political and diplomatic level between the two sides. The Chinese side is willing to push forward bilateral negotiations on the boundary issue and seek fair, reasonable and mutually acceptable solutions in the spirit of peace and friendliness, equal consultation, mutual respect and mutual accommodation. Before the final settlement of the boundary issue, the Chinese side is willing to work together with the Indian side to jointly maintain peace and tranquillity in the China-India border areas." As is evident, the border issue would remain unsettled for a considerable period. The fact remains that China disputes the international boundary between India and China. Indian territory under occupation by China since 1962 comprises approximately 38,000 km. Also, Pakistan has illegally ceded 5,180 sq km of Pakistan Occupied Kashmir (POK) to China in 1963. Further, China is undertaking infrastructural improvements in POK despite the Indian government's requests to cease such activities. Thus, India needs to take concrete steps to measure up to the Chinese stance on the border.

The Parliamentary Standing Committee on Defence, in its report submitted to the Parliament during the monsoon session of 2012, expressed alarm about critical deficiencies in the firepower systems of the three Services. It highlighted the deficiencies of artillery guns, night sights of tanks, air defence weaponry, helicopters of Army Aviation, ammunition holdings and delay in procurements of guns and aircraft. It also mentioned the need for bullet-proof jackets and training aircraft as also simulators for the three Services. All these items are in the process of procurement and require a joint effort by the government and the Services to speed up the process.

The Defence Procurement Procedure (DPP)-2011 gives details of the procurement process for the capital equipment for the three Services. "The objective of this procedure is to ensure expeditious procurement of the approved requirements of the armed forces in terms of capabilities sought and timeframe prescribed by optimally utilising the allocated budgetary resources." The DPP was first published in 2002, revised in 2005, 2006, 2008, 2009, resulting in the current document of 2011.[2] The procurement procedure is structured scientifically to optimise acquisition of equipment in a judicious manner. Currently, the procurement process is being handled by officers from the three Services without any specialised training. During an international seminar on defence acquisitions, eminent speakers brought out that defence acquisition is a cross-disciplinary process needing expertise in technology, military, finance, quality assurance, market research, contract management, project management, administration and policy-making.[3] The US has a Defence Acquisition University established for conducting training courses in 13 career disciplines with three certification levels. The UK and France also lay great emphasis on training their personnel on the subject. There is a need to train officials of the ministry and personnel from the three Services in this field by setting up an institute, as suggested by Maj Gen Mrinal Suman in his article, "Defence Acquisition Institute: A View Point" which appeared in the IDSA, *Journal of Defence Studies*, Vol 6, No 2, April 2012.

The procurement process has comprehensive steps and commences with the formulation of a General Staff Qualitative Requirement (GSQR). Prior to the formation of a GSQR, a Request for Information (RFI) is initiated which gives the market availability of the product. Based on this, the user prepares a GSQR which is ratified by the General Staff Equipment Policy Committee (GSEPC) and the GSQR is finalised. The Defence Acquisition Council (DAC) ratifies the Acceptance of Necessity (AoN), quantity vetting and the approximate price. Based on the directions of the DAC, a Request For Proposal (RFP) which includes all the parameters of the equipment is forwarded to vendors. The vendors respond to the RFP in two parts, the technical part and the commercial part, to the Ministry of Defence. The ministry thereafter technically evaluates the equipment and whereever feasible, trials of the equipment are undertaken. Prior to this, it is preferable that multiple vendors are available for trials to ensure that there is a healthy technical and commercial competition, thereby obtaining a product which has withstood competition. The trials lead to evaluation of the equipment, and the equipment which qualifies in the trials is called for commercial negotiation. The commercial bids are opened and negotiations are undertaken with the lowest bidder. Thereafter, commercial negotiations are undertaken and the negotiated price is vetted by the authorities in the Government of India and approved by the appropriate authority following which a contract is signed, leading to induction of the equipment.

The entire process, undertaken speedily, should be completed in about three years but there are proposals which have been severely delayed, thereby compelling reliance on obsolete equipment which is prone to defects and difficult to maintain. With regard to firepower systems, the cause for delay is due to a single vendor measuring up at the technical evaluation stage or due to non-compliance with the GSQR during trials or finally, the Original Equipment Manufacturer (OEM) not meeting deadlines. The vendors also resort to malpractices which cause the procurement process to be stalled or the vendor to be blacklisted. The Ministry of Defence is leaving no stone unturned to plug the loopholes and progress is being made at a deliberate pace. However, it must be understood that most of the equipment reaches the trials stage but is non-compliant with one or more parameters of the GSQR. The GSQR is a well prepared document based on inputs received in the Request For Imformation (RFI) as also the operational imagination of the user. At this stage, there must be a provision for a collegiate decision to consider the issue and give a waiver to save the proposal from being rescinded, and ensure that valuable time is saved. Further projects on fast track must materialise in time to ensure that operational needs are met.

Currently, numerous firepower systems are under procurement. These include the rifles, carbines and Anti-Tank Guided Missiles (ATGMs) for the infantry, night vision equipment for tanks, the 155mm (52 calibre) howitzers, 155 mm ultra light howitzer,s weapon locating radars and Heron UAVs for the artillery, helicopters for the Army Aviation, a plethora of AD weaponry and radars for the Army Air Defence and numerous combat engineer equipment. As far as the Navy is concerned, there are two aircraft carriers, 46 ships and submarines, 8 additional P-8I maritime reconnaissance aircraft and multi-role helicopters. The aim is to have a 138-ship Navy, with 3 aircraft carriers. The Air Force has the 126 MMRCA Rafale jets, 250 FGFA stealth aircraft, Apache attack helicopters, Chinook heavy lift helicopters and additional heavy lift transport helicopters. The UAVs are with all three Services and there is a requirement of UCAVs in the armed forces as also loitering missiles. Apart from all this, the Services are procuring the BrahMos supersonic cruise missile and ballistic missiles are being procured by the Strategic Forces Command.

We are also developing the 1,000 km Nirbhay subsonic cruise missile whose test flight is to take place shortly. There are going to be developments with regard to MIRV and ASAT missiles. Overall, our procurement process moves at a deliberate pace and the process is affected by turbulence caused by complaints and use of malpractice by some bidders. Accordingly, while we are procuring equipment from abroad, our focus should be on indigenisation.

## Indigenisation

To win wars, a country must be able to stand on its own feet. China has been able to indigenise a fair portion of its weaponry, leading to strengthening of its CNP. Capabilities of the armed forces are measured by the combination of doctrine and technology. In difficult situations, a nation has to stand on its own feet as critical spares and ammunition are denied due to multifarious reasons. To fight through the fog of war, self-reliance is a must and every effort needs be made to make defence indigenisation a success. It is pertinent to add that turning points in history like the break-up of the Soviet Union caused a difficult situation for our country as the spares were not available for 70 percent of our equipment which was of Soviet origin.

It is interesting to note the key facts in the trends in international arms transfers in 2011, as given in the SIPRI Fact Sheet of March 2012. At the outset, the volume of international transfers of major conventional weapons in the period 2007-11 was 24 percent higher than in 2002-06. The five biggest suppliers were the United States, Russia, Germany, France and the United Kingdom. They accounted for 75 percent of the volume of international arms exports. The five biggest recipients were India, South Korea, Pakistan, China and Singapore. Forty-four percent of the imports came to South Asia and Asia-Pacific. This was followed by Europe with 19 percent, the Middle East 17 percent, the Americas 11 percent and Africa 9 percent.

India's imports increased by 38 percent between 2007 to 2011. The primary weapon systems included 120 SU-30s and 16 MiG-29Ks k, both from Russia. The reason for such imports is to obtain state-of-the-art equipment by the Services to ensure effective performance, leading to victory in operations. Accordingly, indigenous products which would be effective on the battlefield will be acceptable to the user. Indigenisation involves the Defence Research and Development Organisation (DRDO), Defence Public Sector Undertakings (DPSUs),

Ordnance Factory Board (OFB) the private sector, the Services and the political leaders. There is a need to find a way by which our indigenisation increases and imports reduce. Currently, about 70 percent of our equipment is imported and there is a need to reduce the percentage of foreign equipment in our inventory.

There is a need for state-of-the-art equipment in combat to win a tactical encounter. Our development agencies cannot produce such weapons with a singular approach. Advanced technologies with regard to defence equipment are available with the US, Russia, UK, France, Sweden and Israel. In the present globalised environment, with less controls, it is possible to obtain transfer of technology with intelligent negotiations. The major issue about technology is its speed of change. Accordingly, there is a need for speedy negotiation and absorption to avoid the product becoming obsolete prior to its introduction. Further investment in modern defence equipment entails heavy initial expenditure with a long gestation period before income starts flowing in. In the Indian environment, government agencies have the capability of investing large amounts for developing a product and private industry has the ability to speedily absorb and produce an item, therefore, a combination of the OEM, the government agency and the private sector would possibly be the pragmatic path to be adopted to enable a high degree of indigenisation.

The concept stated has to be applied in a judicious manner to facilitate the indigenous manufacture of high technology products. The process can be effective only if a level playing field is provided where each agency can contribute effectively, leading to optimisation of resources. The overall coordination of this activity will involve the active participation of the Ministry of Defence. As stated earlier, the Ministry of Defence has set into motion a Defence Procurement Procedure (DPP) which has matured and streamlined the acquisition process considerably. The procedure has a highly structured defence acquisition organisation, with the Raksha Mantri at the apex, a second tier under the Defence Secretary, with the Services represented by the Chiefs at the Defence Acquisition Council and the Vice Chiefs at the Defence Procurement Board. The Defence Acquisition Wing with the Technical Managers from the three Services as well as the Finance Managers from the Defence Finance is the pivot around which the acquisition process rolls. The latest DPP-2011 has added a buy and make (Indian) category which enhances the indigenous capability in acquisition of new products. The Indian company has to absorb 50 percent of the critical technologies, thereby enabling it to manufacture a high-tech product indigenously. The DPP also lays down the need for the vendor to provide offsets for all buy (global) transactions exceeding an amount of Rs 300 crore. Offsets allow spinoffs in multifarious fields. The list of eligible offsets includes civil aerospace, including aircraft with components, a wide range of weapons for counter-terrorism, and training. Our industries will definitely benefit and this will promote further indigenisation

The Indian armed forces would have limited orders. Accordingly, there is a need to promote exports of defence products to generate economies of scale as applicable to mass production. The Group of Ministers has clearly identified the need to expedite issues in this sector.

There are two projects which are role models and are the harbingers for emulation. First of all is the Pinaka multi-barrel rocket launcher. The project was developed by the Armament Research and Development Establishment (Pune) and the launchers are manufactured by Tata and Larsen & Toubro. The ammunition is being manufactured by the Ordnance Factory Board and, as on date, two regiments have been operationalised. It is pertinent to note that two more regiments are being raised shortly.

The best example is the supersonic cruise missile, the BrahMos. This is a joint venture between India and the Russian

Federation. The Indian contribution is 50.5 percent and the Russian contribution is 49.5 percent. The missile is presently being manufactured jointly and gradually all the components are being indigenised. While the BrahMos is an Indo-Russian joint venture, a number of industries are involved in manufacturing the components. The airframe is being made by Godrej, the composites are being made by Larsen&Toubro, the inertial navigation system is being made by HAL and the onboard computer system by Datapatterns. As on date, the weapon system has been inducted into the Indian Navy and Indian Army. Recently, the missile has proved its steep dive capability, enabling it to be deployed in the mountains. The air-launched and submarine-launched versions are under development and will undergo evaluation shortly. Overall, the process is successful and deserves to be emulated.

The synergy among the user, the developer and the manufacturer, with suitable coordination by the Ministry of Defence will pave the way for indigenisation of defence products. Further, in order to reach economies of quantity, exports must be permitted of selective items. Another interesting case is the indigenous development of the 155mm (45calibre) gun by the OFB. The OFB had received the ToT documents from the original equipment manufacturer, BAE Systems, when the guns were procured in 1986-87. The Indian Army has been correct in requesting the OFB to develop two prototypes and thereafter go in for a bigger order, to start with, possibly, about 10 regiments, thereby enhance the order based on the requirement. The prototypes are shaping up well and the success of this project will be an important milestone in the modernisation process. Similar efforts are being made to operationalise the nuclear submarine the *Arihant* and the indigenous aircraft carrier, the INS *Vikrant*. The fighter aircraft Tejas is in the final stages of acceptance and the UAV Nishant has proved its mettle and we are going ahead with the development of the Medium Altitude Long Endurance (MALE) UAV Rustam as also the UCAV Aura.

In the field of SAMs, there is a joint venture between DRDO and IAI, Malat, for the 70 km Barak NG missile and another joint venture with the same company for the 140-km LRSAM for the Air Force and Navy. In the field of anti-ballistic missile systems, we have successfully tested the PAD missile for high altitude interception and AAD for low altitude interception. Further, two new ballistic missiles interceptors, the AD-1 and AD-2, flying at hypersonic speeds are being developed to intercept missiles at a range of 5,000 km. Efforts are also being made to import the ABM system Aegis from the US or the Arrow from Israel.

## Long-Term Integrated Perspective Plan (LTIPP)
The Kargil conflict of 1999 led to the formation of the Kargil Review Committee (KRC). The KRC presented its findings to the government and these were tabled in the Parliament in the beginning of 2000. The Prime Minister, based on the report, set up a Group of Ministers (GoM). This is normally referred to as the Arun Singh Committee, which was tasked to examine the report holistically. The GoM realised that a lot of improvement was needed in the integration of the three Services. They gave numerous recommendations. A major recommendation was the creation of the Integrated Defence Staff (IDS). The group strongly recommended that there is a need to integrate the process of planning for achieving synergy in the perspective planning and implementation. Accordingly, they stated the need for a Chief of Defence Staff (CDS) and the necessity of a new headquarters of the IDS. Accordingly, a new Headquarters IDS was raised on October 01, 2001.

The IDS consists of branches comprising Operations, Doctrine Organisation and Training, Intelligence and Policy Planning and Force Development (PP&FD). The PP&FD branch is responsible for the process of generating the Service plans. The major duties consist of force structure development, budget analysis, acquisition, procurement and technology management, formulation of the Long-Term Integrated Perspective Plan (LTIPP) and Five-Year Defence Plans, proposing prioritisation of schemes, coordinating strategic and security perspectives, analysing critical deficiencies in force capabilities and assessing the impact on objectives important to the nation.

The production of the LTIPP is a comprehensive task involving collaboration with multiple agencies. These are the Cabinet Committee on Security (CCS), National Security Adviser (NSA), Ministry of Defence (MoD), Ministry of Home Affairs, Ministry of External Affairs (MEA), Ministry of Finance (MoF), intelligence agencies, HQ IDS, Army HQ, Naval HQ, IAF HQ, DRDO, and industry to include Public Sector Units (PSUs) and the private sector. The procedure entails five stages. The first stage comprises articulation of a national security strategy. This would be conceptualised by the CCS, assisted by the NSA. The second stage would be formulation of defence planning guidelines visualised by the MoD and these would state the contingencies the Services would be expected to respond to in the next 15 years, The third stage would be formulation of a defence capability strategy. This would be undertaken by the IDS and three Services.

This document would analyse the capabilities needed for these contingencies and assess the capabilities required, with the existing capability, and the *modus operandi* needed for filling the gaps with priorities. The fourth strategy would be preparation of a Defence Capability Plan. This would be prepared by HQ IDS in consonance with the Services. This would have a span of 15 years and would list the capabilities with associated timeframes; also how these options would be achieved, whether by indigenous development or by agencies from abroad and the broad nature of each project. The fifth stage would be the preparation of the LTIPP.

The LTIPP would be the extension of the Defence Capability Plan and would contain the programmes and projects required to be undertaken to reach the targets of the Defence Capability Plan. The present LTIPP was prepared by HQ IDS after receipt of the Long-Term Perspective Plans from the three Services. This entailed study of force levels, force structures, accretion and inter-Service prioritisation. The HQ IDS has to get the Chiefs of Staff Committee (COSC) to vet and approve the LTIPP. Thereafter, it is a steep climb to the Defence Planning Council and, finally, the approval of the CCS. It is, indeed, a complex process and entails going into explicit details with military precision.

The LTIPP, from 2012 to 2027, was approved by the Defence Acquisition Council on April 02, 1912. This should have been approved in the year 2011 but was delayed by a year. The 12th Five-Year Plan, from 2012 to 2017 was also approved on the same day.

Consequent to the clearance which covers the vision for the 12th, 13th and 14th Plans, the unclassified version of the LTIPP will be promulgated in the form of a Technology Perspective Capability Roadmap (TPCRM) to enable the DRDO, defence PSUs and private industry to plan their research and development roadmap.

It was also observed during the meeting that the projects take a long time to fructify. The Ministry of Defence would like the Services to ensure that slippages do not take place. The Defence Minister further clarified in Parliament that the defence expenditure for 2012-13 would be $ 74.5 billion. The IAF would expend $ 30.5 billion, the Navy $ 24.7 billion and the Army $ 19.3 billion. The approval of the LTIPP is a positive step towards modernisation of our armed forces. Time-bound implementation would result in a Revolution in Military Affairs (RMA) in the three Services which is needed to face our adversaries effectively.

## Likely Timelines for Induction of Firepower Resources: A Reality Check

Acquisition of equipment, resulting in definite timelines, is extremely difficult in our country. We have had long gestation periods for all firepower resources in the three Services. Accordingly it is worth doing a reality check. There is a need for leaders, bureaucrats and soldiers who would like to change gears, accelerate and reach their destination with speed and precision in terms of time. To quote a modern adage, "The people who are crazy enough to think they can change the world are the ones who do."[4] Do we really have diehard, crazy acquisition teams who can revolutionise our three Services by getting their equipment inducted in a reasonable timeframe? The answer is that it is an uphill task to which we can give our best and thereby achieve pragmatic targets.

Firepower in the Indian context has $C^4I^2SR$, weapon platforms, ammunition and Post Strike Damage Assessment (PSDA). In as much as $C^4I^2SR$ is concerned, to be effective, it must be integrated with sensors, staff and communications. Integration among the Services is being done at a deliberate pace with each Service taking baby steps and ISRO gradually providing them satellites, and procurement being done of UAVs for catapult and runway take-off as also rotary-wing UAVs for the Navy. An integrated system should take a minimum of 12 to 15 years, and by 2030, we would be somewhere with an integrated situational awareness system. This would be a big step, as this would resolve issues pertaining to targeting and allotment of resources.

Weapon platforms and ammunition form the next link in the firepower chain. As regards, the Indian Army, we will start with small arms. The modernised assault rifle, the carbine and the spike or the Javelin ATGM should be inducted, possibly by the end of the 12th Plan, and the air burst grenade, mini UAVs and a digitised battlefield management system by the middle of the 13th

Plan. Further, mine protected shoes, light bullet-proof gear and state-of-the-art night fighting equipment should be provided by the end of the current Five-Year Plan, around 2017.

The regiment of artillery will be the battle winning factor in any future war. It is a proven fact that effective artillery fire broke the will of the enemy to fight during the Kargil conflict of 1999, resulting in the recapture of Tololing and Tiger Hill. It is pertinent to note that despite the KRC recommendations, modernisation of the arm is moving forward at a slow pace. Induction of Smerch rocket launchers, Pinaka rocket system, extended range rocket ammunition, BrahMos supersonic cruise missile, upgrading of 10 regiments of the 130 mm to 155mm (45 calibre) guns, AN TPQ-37 weapon locating radars, LORROS, BFSR ELM 2130, Heron and Searcher UAVs as also Artillery Combat Command and Control System (ACCCS) have certainly added punch to the predominant arm that will provide firepower. The obvious question is: what is the way ahead? The answer is complex as the procurement of artillery has been a difficult task. At the current juncture, the first to be inducted is likely to be the 155 mm ultra light howitzer. Induction of this weapon system by the Foreign Military Sales (FMS) route should commence by 2015 and the complete package of eight regiments would possibly be inducted by 2020. The gun is proven equipment in the battlefields of Afghanistan and there is a possibility of a repeat order, resulting in an additional eight regiments. The next equipment would be the 155 mm (45 calibre) being manufactured by the OFB. The user evaluation is being processed and the induction is likely to commence by 2016-17. About 10 regiments are likely to be fully inducted by the middle of the 13th Plan. The 155 mm tracked SP gun is currently under technical evaluation and possibly after trials, will be inducted around the middle of the 13th Plan. The 155 mm (towed gun), 155 mm wheeled SP, 155 mm mounted gun system and the 120 mm mortar are in their initial stages for procurement, and if they qualify in the evaluation trials, induction would commence possibly five years after the trials. The BrahMos steep dive missile is in the final stage of procurement and induction would possibly commence by 2014. Two additional regiments of the Pinaka are likely to commence induction by about the middle of the 12th Plan. The indigenous Weapon Locating Radar (WLR) is on user evaluation and induction would commence in about three years. Loitering missiles and UCAVs would possibly be inducted by the end of the 12th Plan. Induction of this wide array of equipment will add punch and vigour to the artillery of the Indian Army, matching it qualitatively with the Chinese and Pakistani artillery. Of course, the triad of the ministry, DRDO and regiment of artillery has to synergise its actions.

The mechanised forces are enhancing the numbers of T-90 and upgrading the T-72 with superior night fighting equipment, and ammunition is being indented for operational use. The Spider or the Javelin will be inducted, possibly by the end of the 12th Plan. The indigenously developed Nag will have to improve the hit probability in higher temperatures which the DRDL scientists of DRDO are working on to develop and it would possibly take about a year or two before the fire and forget missile becomes compliant. It is encouraging to learn that mechanised forces are going to be deployed in high altitude regions and would be employed in selected areas. Two armoured brigades with T-90 tanks and the BMP-2 APC are likely to be deployed—one in eastern Ladakh and the other in Sikkim, giving us the punch needed in these mountainous areas. The Army Air Defence needs to be provided with a modern set of missiles and radars. Our current systems need to be upgraded or replaced. The old faithful L-70 weapon system needs to be upgraded and our missile systems also need to be replaced. This is applicable to the Very Short Range Air Defence (VSHORAD) system and long range system. It is, indeed, heartening to see the Akash, which has a range of about 35 km, being inducted The other firepower assets need to be expedited to meet the challenges.

The Indian Navy has commissioned 15 ships in the last three years. These comprise three Shivalik class stealth frigates, *Shivalik; Satpura and Sahyadri*, two fleet tankers *Deepak* and *Shakti*, one follow-on 1135.6 class stealth frigate, the INS *Teg*, the sail training ship *Sudarshini* and eight water jet fast attack craft. The INS *Chakra*, the nuclear attack submarine, was commissioned on January 23, 2012, and our Navy is among the six countries that operate nuclear submarines.[5]

The future procurements of the Indian Navy would enable the Service to perform its firepower tasks effectively and stand up to any opposition from our adversaries. The warships likely to be inducted are the aircraft carrier *Gorshkov* (*Vikramaditya*), currently on trials in the White Sea, with delivery slated to be in early 2013, and the air defence ship (*Vikrant*) progressing at Cochin Shipyard Ltd; with delivery by 2017; two stealth frigates on similar lines as the INS *Teg*, with the BrahMos missiles, and Shtil anti-aircraft guns by early 2013. Seven Type 17-A stealth frigates are under construction; four at Mazagaon Docks Ltd (MDL), Mumbai, and three at Garden Reach Shipbuilders and Engineers (GRSE), Kolkata. In addition, three Type-15 A destroyers, an improved design of INS *Delhi*, fitted with the M-star radar, the BrahMos missiles, long range Baraks and two multi-role helicopters are being manufactured by MDL in a joint venture being processed between MDL and Pipavav Offshore and Engineering Services Ltd. Further, six Scorpene submarines by DCNS Navantia in combination with MDL are being processed. The most important development is in the field of the indigenous nuclear submarine, with the INS *Arihant* likely to proceed for sea trials and to be fitted with K-15 nuclear capable missiles with a range of 700 km. It is reported that two more nuclear submarines, the S-4 and S-5, are under construction at the ship-building centre at

Vishakapatnam. For maritime reconnaissance to gather inputs in selecting targets, four Boeing P-8Is are to be inducted from 2013 onwards.[6]

The modernisation of the Indian Navy is proceeding on an even keel. With its acquisition of a variety of warships, it would be able to maintain ascendancy of firepower from the Persian Gulf to the Strait of Malacca. Vis-a-vis China, which is on the way to induct an aircraft carrier, our Navy would be able to undertake its task with elan, pride and military precision. The firepower assets have matured and it would be possible to synergise and control the high seas in our area of responsibility.

The IAF has commenced its modernisation process with the aim of building itself into a future aerospace power. The transformation of the IAF is three-pronged: the first is to maintain and preserve existing firepower assets, the next is to selectively upgrade the existing fleet, and the third is to procure state-of-the-art aircraft. All this leads to improvement of firepower. As far as the upgrades are concerned, these entail upgrading of the MiG-29 by 2017. The Mi-17 helicopter and SU-30 are already under progress. Further modernisation of airfield infrastructure is also being progressed.

As regards procurement, the commercial negotiations are on for procurement of 126 MMRCA with Rafale. Additional SU-30s are being acquired. Induction of the LCA is to start shortly. All six AWACS aircraft are under evaluation. Further, the FGFA are slated to be available by the end of the 12th Plan and induction will be complete of 250 aircraft by the end of the 14th Plan. All these, along with the new generation of transport aircraft, Hercules and Globemaster, will add to our firepower from the air. The air-launched version of the BrahMos is under development and will definitely be inducted by 2015. If things move as per the stated timelines, we would be able to definitely combat the PLAAF in a full spectrum conflict over Tibet.[7]

As regards missiles, we would be shortly testing our subsonic cruise missile Nirbhay, slated to have a range of 1,000 km. The induction of the missile would be based on the progress of trials. Normally, a missile is put through six flights before user acceptance.

However, in a full spectrum conventional conflict, we need to engage targets at ranges between 600 to 1,000 km and this entails the need for speedy development. Further, we need to develop ASAT weapons, ICBMs with MIRV as also MARV capabilities and a credible multi-layered ABM system capable of countering the Chinese and Pakistani missiles. We would also need our own navigation satellites to navigate these missiles.

**The Way Ahead**

Asymmetry of firepower will lead to victory in the 21$^{st}$ century. India must prepare itself militarily to deal with a two-front war. The adversaries are well equipped to fight a digitised battle in the 21$^{st}$ century. A ground attack would be preceded by cyber attacks, destruction of Indian satellites, concentrated missile attacks on selected communication nodes, integrated electronic attacks to paralyse C$^4$I$^2$SR networks and, finally, an air-land offensive with objectives in Aksai Chin and Arunachal. Shallow penetration would be undertaken all along the LAC.

India needs to, firstly, have adequate force levels to dissuade our adversaries from undertaking an offensive. While two divisions with their artillery complement have been raised, there is a need to raise a strike corps for the mountains. This strike corps, apart from two divisions, must have an artillery division to provide the requisite firepower to this mountain spearhead. The Air Force must deploy additional squadrons to effectively deal with the PLAAF as also the assault columns and their logistics echelons. The Navy must get the nuclear submarines operationalised to ensure that the nuclear triad is complete. Above all, the three Services must fight an integrated battle with C⁴I²SR capabilities. The Naresh Chandra Committee has submitted its report and the same should be pragmatically implemented.

Specific to firepower on land, we must speedily acquire eight regiments of the 155mm (calibre) ultra light howitzers along with Chinook heavy lift helicopters. For the guns, if found suitable, a repeat order could be given and that would provide credible firepower to the mountainous areas. The 155 mm (45 calibre) being built by the OFB must be expedited and 10 regiments inducted by the 13th Plan. The procurement process particularly for the 155 mm (52 calibre) towed gun and 155 mm (52 calibre) mounted gun system must be expedited. The Smerch rocket systems is an excellent weapon system, needed for the mountains. We need additional UAV troops as also UCAVs at the rate of a battery per division to effectively counter the adversary's offensive. Further, loitering missiles would pay rich dividends. Our present set of weapon locating radars are not suitable for the mountains and, if required, we must obtain suitable equipment for this purpose. Firepower elements need roads to reach their designated areas of battle. The Border Roads Organisation must be able to complete the roads on the border speedily. Further, as per the latest directions of the Ministry of Defence, the Indian Army will soon have attack helicopters inducted for primarily operating with the three strike corps. It is obvious that the future strike corps in the mountains would be equipped with suitable attack helicopters for operating in the regions bordering China.

The Air Force procurements of 126 Rafale MMRCA fighters must be completed at the earliest. The FGFA T-50 must commence induction in the 13th Plan and the LCA must be operationalised. The Apache attack helicopters and the Chinook heavy helicopters must be acquired, inducted and fine-tuned with the Army.

The Navy must be able to launch ballistic missiles from nuclear submarines. The BrahMos supersonic cruise missile with steep dive capability must be inducted in greater numbers and the air version developed to ensure deep engagement of land targets. DRDO must focus on the subsonic cruise missiles, ICBMs with MIRV and MARV capabilities as also a credible multi-layered ABM systems and directed energy weapons.

To conclude, firepower in 2030 will be the predominant factor that would determine the fighting capability of a nation. Victory can be assured only by building asymmetries of firepower. There would be a need to synergise firepower at a tri-Service level to pave the way for success. To overcome the challenges of technology, a triad needs to be formed among the user, the developer and the manufacturer. In the Indian perspective, the three Services must speed up the modernisation process. The present set-up of obsolete weaponry will be of little use in a two-front war. In a network-centric environment, a real-time link among the sensor, command elements and the shooter is required. It is extremely important that each divisional artillery brigade of the Army has a battery of UCAVs. Precision would be necessary for all engagements by platforms. Our present holdings are extremely low. There is an urgent requirement to ensure that 50 percent of our ammunition should be PGMs by 2030. We must also develop rocket ammunition to enhance the range of the Pinaka to 60 km and of the Smerch to 120 km. Further, our ballistic missiles must range 10,000 km with MIRV and the BrahMos cruise missile must attain hypersonic speed. By 2030, weapons in our inventory would be able to simultaneously engage targets in the depth, intermediate and contact battle. In order to exercise judicious control over all platforms, there is a need to have an artillery division at the corps level to effectively execute joint fires. In all this, a sense of urgency must be felt by the Services and all agencies to expedite the procurement process and change organisations to revolutionise our firepower capabilities by

2030. In a two- front war, the three Services need to be integrated to dissuade our adversaries from fighting a war.

To simplify the aspects, a table is indicated below showing the weapon systems currently held, with likely weaponry in 2030. The table covers some of the weapon systems of the Indian Army and their associated technologies. Details and the connected issues have been explained in the monograph. Details of the Navy and Air Force have been covered comprehensively.

**Table 1**

| FIREPOWER ASSETS | 2012 | 2030 | REMARKS |
|---|---|---|---|
| Strike Corps (Mountains) | One possibly by 2017 | Additional for Ladakh and J&K | |
| UCAV battery | Nil | Need for one per division | Case for consideration |
| Loitering missile | Possible induction by 2017 | Need for one per corps | RFP issued, needs to be expedited |
| Heron UAV | Two troops and two troops awaiting contract signing | Need for similar UAV -10 troops | Rustam being produced by DRDO, needs to be expedited |
| 155 mm (39 cal) ULH | LOR issued to the US government Induction of eight regiments likely to commence in 2017 | Eight regiments likely to be inducted and there is a possibility of a repeat order | Proven gun in Afghanistan |
| 155mm (45 cal) | Trials being undertaken by the Indian Army. | About 200 guns likely to be inducted | Being manufactured by OFB, Jubbalpore |
| 155 mm (52 cal) Towed gun | Technical Evolution Committee (TEC) being undertaken | Possible induction | Trials would indicate further progress |
| 155 mm (52 cal) MGS | RFP being issued | Likelihood of induction | |
| 155mm (52 cal) Wheeled (SP) | RFP being issued | Possibility of being inducted | |

| 155 mm (52 cal) Tracked (SP) | TEC being undertaken | Possible induction | Trials would indicate further progress |
|---|---|---|---|
| Smerch for mountains and 120 km | | | Under consideration |
| Long range mortar | RFP being prepared | Likely to be inducted | |

## Notes

1.  "China's Strategic Missile Realise Mobile Launch," http:/ www.globaltimes. cn/content/730540.html .
2.  "Defence Procurement Procedure-2011, Capital Procurement", Ministry of Defence, Government of India, 2011.
3.  See    http:/business-standard.com/india/news/cagdefence-buys-tehelka- syndrome/285130 and article by Maj Gen Mrinal Suman, "Defence Acquisition Institute a View Point,", in *Journal of Defence Studies*, Vol No 6, April 2012 (New Delhi: Institute for Defence Studies & Analyses).
4.  Walter Issacson, "Apple's Think Different Commercial in 1997,"in the book *Steve Jobs*, page prior to contents, published by Gopson Papers Limited, Hachette India, Noida.
5.  Adm Nirmal Varma (Retd), "Indian Navy's Recent Milestones,"excerpted from the Chief of Naval Staff's farewell statement at the press conference on August 07, 2012. Printed in *India Strategic*, Vol 7, Issue 8, August 2012, pp 14-15.
6.  "Current Order Book of Indian Navy", *India Strategic*, Vol 7, Issue 8, August 2012, p. 19.
7.  Air Chief Mshl P V Naik (Retd), " The Indian Air Force and 24x7 capabilities", *India Strategic*, Vol 7, Issue 7, July 2012, pp. 10-11.

www.ingramcontent.com/pod-product-compliance
Lightning Source LLC
Chambersburg PA
CBHW020209290326
41948CB00002B/139